BOXED & LABELLED TWO!

✶ NEW APPROACHES TO PACKAGING DESIGN ✶

gestalten

BOXED & LABELLED TWO!

TABLE OF CONTENTS

pivo

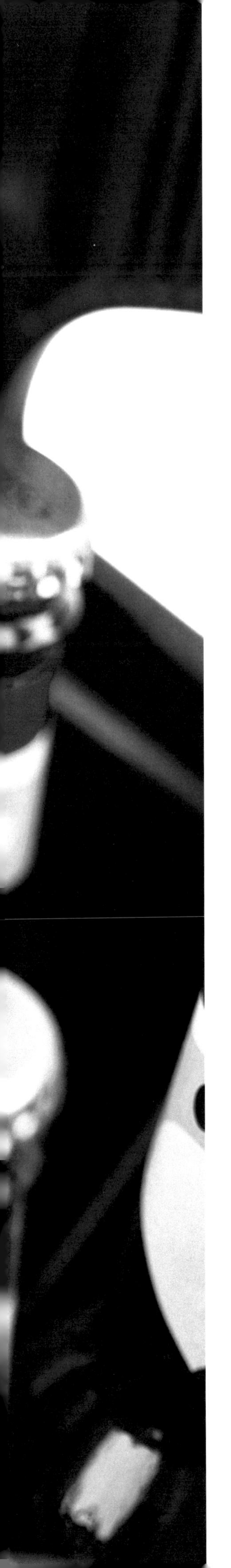

Watching my six-year-old son as he unpacked his birthday present the other day, it struck me how this simple event contained all the ingredients needed for a winning package design, but that it also represented many major environmental pitfalls. It was a highly emotional experience shaped by desire, restlessness, excitement, elation, emotion, joy, and possession. Packaging embodies such magic and fascination. Its central element is the pleasure of the conquest; all formal, functional, and aesthetic qualities cannot compete with this notion of the pleasure associated with the unpacking experience, which reaches its ultimate climax in the attainment of the object of desire. The excess of colored papers and ribbons, which exercised so much fascination and excitement only a few seconds earlier, lose all their appeal in an instant and are then thrown away in a pile of unwanted debris. Herein lies the very paradox of packaging; it is an accessory to a desire that once satiated transforms the package into superfluous waste that must be quickly disposed. How can we manage all the multi-matter trash?

UNPACKING

The vast majority of consumer products could not exist without a specific package. Like the skin of a banana, packaging is an integral and inseparable part of the product. So what then, is a good packaging design? In addition to its strict function of transporting a product from point A to point B, are there measurable qualities to assess the features of a winning design? Most of all, packaging design is a challenge because it must reconcile the expectations of many stakeholders. The producer wants the product delivered to its destination in the best possible condition for its preservation. As such, the evaluation will be based mainly on functional criteria. The merchant hopes that the brand will capture attention, stand out on the shelf, and ultimately purchased. And the consumer demands maximum service at a minimum price. Nevertheless, none of these criteria alone or even in combination will ensure the success of a brand. So what, then is the magic that will lead to the creation of icons such as the Heinz ketchup bottle or the Chanel No.5 bottle? Those that would claim to know the secret are very wise indeed, but at the very least we can say that design plays a central role in the equation.

A COMMUNICATION MEDIA

Since the beginning of the industrial era, packaging has gradually replaced the retailer by establishing itself as the primary interface between the producer and the user. Completely embedded within the industry, it underwrites all aspects of the sale. It conditions, gathers, protects, seduces, values, informs, advises and assists the user from the shelf to the cash register, and finally to the consumption of the product itself. It is the ultimate tool of counter-service. Of all communications media packaging is among the most ubiquitous—it is able to establish a tacit relationship with the user thereby accompanying the user in their daily actions, from the most intimate to the most mundane.

But not all packages have the same direct relationship with the user. Some packaging components provide a purely functional role in the protection and organization to assist with delivery of the product. These secondary boxes and plastic wrappings whose function is to protect the product during transport do not have the same impact on the sale as does the packaging that directly interacts with users on the shelf. Often, there are additional boxes or structures that are designed to support the branded product— these too play a strategic role in the business of seduction because the opportunity to attract, charm, and convince the user lasts only a split second. Finally, there is another type of packaging that is part of the product itself and which assists the user in its application: soft drink bottles, tubes of toothpaste, and honey bears all become the medium through which brands deliver their product and their promise.

Given its strategic position in the consumer's life, packaging has become an almost perfect reflection of the consumer's lifestyle and values. Packaging speaks to us as consumers, to our desires, to our guilty pleasures, and to all the elements that comprise our humanity. It tells the story of a product in which we play the decisive role, particularly at the moment of purchase.

LOVE AT FIRST SIGHT

It is a story of seduction between the product and the user. To appeal, packaging must deal with two contradictory principles; it must distinguish itself yet prove that it belongs to a class of recognizable products. If the packaging is too distinctive, it may be overlooked, but if it is too common, it may also go unnoticed. The art of seduction lies somewhere between these two absolutes. What governs the rules of seduction between people—namely beauty, elegance, distinction, humor, fun, intelligence, altruism, and envy—also prove effective as means to attracting the consumer. The package must have its own personality and tell an original story that must hold the attention of users and submerge them in a narrative, hence the importance of not always repeating the same tale.

However, the chameleon effect is observable in many product categories. If you go to the nearest supermarket and take a look at the display of regular fruit juice cartons, you will notice that within the same category of products, most brands tell more or less the same story. In terms of seduction codes, they all utilize the same type of illustrations showing juicy fruits in a whirlwind of saturated colors and fluid fonts. If by chance a brand emerges or changes its packaging shape, its competitors will be quick to follow. "Not very distinctive" you might say? Unfortunately, in such a context, the main criteria for creating distinction will often come down to the price.

THE DESIGN EFFECT

How then can we reconcile the goals of visibility and the producer's profitability with the needs of the consumer? It is the task of the designer to consider how to establish a balanced relationship between these objectives that seem a priori to oppose one another. If the distinction of the brand is the central concern—particularly from a marketing point of view—then the product itself and the customer's expectations are often overlooked. The development of a brand must go beyond mere perception and aim towards the creation of a positive and practical experience for the user. In fact, a brand may very well stand out for all the wrong reasons. A package that is difficult to open, an oversized box or a fraudulent and idealized representation can have a devastating effect on the brand perception. In Canada organic eggs are sold in plastic containers whereas conventional eggs come in a recycled carton container. This is a nonsensical design decision is it not? The progressive health benefits are annulled by the environmental drawbacks of the packaging.

For the buyer, packaging is primarily a tool that provides the best service at a low cost. For an equivalent product, how can a specific brand transcend its primary function and offer that small difference, a piece of information or distinctive service that will make it sell? Increasingly conscious of diet, the environment, and other social issues, today's consumers are in search of truth. They demand more honesty and a coherent discourse rather than hype and pretension. In response to growing public awareness, a proliferation of health, environmental and social issue claims have flooded the market for many years now. These claims increasingly occupy a greater surface on the packaging and encompass a range of issues such as organic and fair-trade certification, Genetically Modified Organism (GMO) certification, or if a paper product is approved by the Forest Stewardship Council (FSC) or to what extent the packaging itself is recyclable or made from recycled materials. These assertions are sometimes standardized, but more often than not they are unverifiable self-declarations. Such information has become so present that it will often overshadow the brand. While in some cases these claims may be credible, they can nevertheless provoke a kind of cynicism in consumers who find themselves lost in a sea of information. Saying too much is sometimes worse than saying nothing at all. Consumers are in search of truth and it is becoming increasingly risky to hand them half-truths or lies. Take for example the ongoing and questionable practice of downsizing in which product quantity is gradually reduced to keep pricing stable. It is a lie by omission. Customers who see through this scheme feel cheated and this practice can have a catastrophic consequence on the credibility of a brand. I am reminded of a biscuit producer who, after I had complimented him on the excellence of his packaging, replied: "Packaging can only ensure the first sale. If there is too much of a gap between the claim and the actual experience, then the story of seduction ends there."

In the end, the best way to create a winning package is not to design it for the consumer, but rather to design it for the user. If the consumer is the one who buys the package, it is the user who uses it. This distinction marks a critical nuance in establishing a dialogue with the client. It reflects a process in which the sale is not a singular goal, but rather one aspect in the overall commercial experience; an experience that includes both the purchase and use of the product as well as the eventual disposal of the product and its packaging. Too many packages are designed exclusively for the needs of the producer at the expense of the user. At the risk of being labeled an idealist, I think the notion of honesty or "truth of the brand" should be a founding principle for the sustainable development of future brands.

Good packaging is not simply a box that has been covered with branding. Rather, it is a much richer outcome that is the result of a collaboration between form, function, and message. Packaging design is a discipline whose process involves a series of decisions that often arise one from another. This interconnectivity requires that the different stakeholders become involved early on in the design process so that the chances of success are increased.

INNOVATIONS

I teach to a new generation of young graphic designers who are passionate about packaging design because in it they see a tangible and contemporary form of communication, one which is closely connected to the consumptive lifestyle in which they have been raised. Critical and sometimes cynical towards consumer culture, they bring a more whimsical and humanist approach to the trade. In my view, a critical approach provides a strong basis for innovation. Before you can suggest new avenues in packaging design, you must be able to deconstruct old habits and preconceived ideas. For example, the fact that this new generation has been eco-conscience since childhood allows them to almost intuitively integrate sustainable concepts into their designs. More than ever, these young designers are freely exploring various types of packaging. They feel increasingly free to borrow techniques and materials from other product sectors. Wine in cartons, milk in stand-up pouches, and spices in test tubes are only a few examples of this tendency to transpose methods and codes. These deviations open the way for innovations that through their originality or their functional relevance allows for the emergence of new ideas.

Historically, significant breakthroughs in packaging have always been intimately linked to technological innovations and to the discovery of new materials. The development of new plastics, for example, radically changed the practices of the 20th century and encouraged the proliferation of new formats and individual portions. More recently, environmental concerns have fostered a plethora of new materials and new methods for packaging. The arrival of bioplastics will certainly make its mark on the packaging industry and will significantly modify the management of waste. Likewise, research is underway to develop replacements for polystyrene based on spongy organic fungus-based life forms. In the coming years, packaging will also be marked by the development of nanotechnology which will, among other things, allow for better real-time quality control of perishable products. Although these technologies may in some cases increase the environmental footprint of certain packaging, they will nevertheless allow us to optimize the distribution of perishable products and reduce losses.

Design is a problem solving process. It is in situations that involve multiple constraints that we can appreciate most how design creativity can be put to use. Innovation often requires a questioning of current practices. Sometimes you have to question everything, and break out of the comfort zone in order to be able to explore new avenues.

Packaging design is currently a booming and popular subject. It is severely criticized by some and adored by others. It represents the best and worst of modern commerce. We live at a time in which the world produces more packaging than ever before, but the pressure to reduce our impact has never been greater. In a globalized market, brands are increasingly looking to revive historical and tribal cultures. Influences are arriving from all directions and graphic codes seem to have no boundaries. While globalization may be worrisome it allows for a remarkable exchange of new ideas and technologies in addition to new technical knowledge utilizing the Web, specialized social networks, and blogs.

The main theme of this book is creativity that is expressed through projects that push beyond preconceptions and clichés. It is an ode to daring and innovation in packaging design. The concepts shown here explore the boundaries of traditional packaging and propose new avenues for shape, material, function, and communication. Through the quality of work presented, this book brings packaging forward as a separate and distinct design discipline. In a world that is getting smaller and smaller, the challenges for tomorrow's packaging designer will be both exciting and complex. Will the new generation be able to find ways of fostering responsible values within commerce? Will the packages of the future preserve that magic for the six-year-old child without ransoming our planet's resources?

MILKSHAKE WITH
PURCHASE OF THE
HANDBURGER ORIGINAL
RAFFLES CITY SHOPPING CENTRE, B1-77/78
FOR DINE-IN ONLY. OFFER VALID TILL 13TH SEPTEMBER 2009.
NOT VALID WITH OTHER PROMOTIONAL OFFERS. WHILE STOCKS LAST.
WE RESERVE THE RIGHT TO AMEND THE TERMS OF THE PROMOTION AT ANY TIME WITHOUT PRIOR NOTICE.
THE HANDBURGER
THE HANDBURGER
THE HANDBURGER
MENU
THE HANDBURGER ORIGINAL $13.8
TEA-SMOKED DUCK BURGER $15.8
CHICKEN CAESAR BURGER $12.8
TANDOOR CHICKEN BURGER $13.8
PULLED PORK BURGER $11.8
PARMESAN PRAWNCAKE BURGER $17.8
BEER-BATTERED DORY BURGER $14.8
STUFFED PORTOBELLO BURGER $13.8
VEGETARIAN KAKIAGE BURGER $10.8
THE WORKS BURGER $16.8
EVERYTHING ELSE
DRINKS AND DESSERTS
THE HANDBURGER
THE HANDBURGER

CHAPTER 1

POP

The sixties rocked my childhood in North America and no packaging better reflects this period than the bottle of Mr. Bubble soap, which accompanied me in the bathroom and stood out from all the other brands by way of its comic and bright pink mascot. More than any other product, Mr. Bubble called to me, speaking directly to my childhood desires and imagination. It did not speak about what I had to do, but about what I wanted to do. It made a promise that bath time would no longer be an ordeal, but instead a party! I do not remember the feel or smell of it, but even when my mother reused the same bottle with a substituted cheaper soap, I could not tell the difference and in this way, Mr. Bubble had a permanent place in our bathroom for years. The packaging was the product and induced an experience of fun and humor. As I could not at the time read English, it took me several years to comprehend the meaning of the words "piggy bank" written on the Mr. Bubble bottle. Only then, did I discover the use for the coin slot on the lid: to give Mr. Bubble a second life as a piggy bank.

The pop style celebrates the package as object. Each surface and function contributes to the telling of a story, but also to the involvement of the user in the scenario. Its visual language speaks freely and outside of the preconceived codes of its time. It stands out by way of its unique and incomparable personality. Contrary to what our parents may have taught us, pop packaging invites to play with our food and to step outside of taboos and censorship. It is festive and exerts a fascination on the user.

By introducing a playful and simple message, pop packaging establishes a direct dialogue. The vibrant colors, the narrative graphic codes, the fun shapes of the containers, all characterize this approach. It celebrates the candor of the child we were and that we might still be. These packaging designs take on the shape of human forms or of everyday objects.

Here packaging is an interactive experience; the chocolate can be a computer keyboard or a Scrabble game; the box of tissues is a slice of fruit; the bottle becomes a sparkling fire extinguisher; the box is divided into a polyhedron set of blocks. The packaging is participatory. Sometimes it is a collectible and is the icon for an entire generation. The pop style in its relation to popular culture speaks about who we are. It brings us back to the here and now. It strives to be a reflection of our culture. It makes us smile and reminds us of our childhood pleasures when we visited the corner store for a treat. All these goodies, like the PEZ dispensers, are emblazoned with the faces of our heroes such as Batman, Buzz Light Year, and Wall-e. The small Bazooka Joe comics included with the chewing gum, candy necklaces, Popeye candy cigarettes, and Kinder eggs—all these packages go beyond the mere role of container and engage the user in a value-added experience. Sometimes the shape of the product itself contributes to its playfulness. *The Obamitas* chocolate cookies with their President Obama likeness or the penguin shaped *Birdy Juice Box*—invite us to transcend the banality of the products we consume. *The Deli Garage* products are also remarkable in the way they relate to the brand's theme in the form of bolts, screws, and glue tubes. With pop packaging, the brand is often fully embodied as packaging adopts a first-person narrative means of communication. Pop packaging expresses an assumed pleasure linked to the act of consumption, which it celebrates with unpretentious lightness.

Pop style packaging is the antithesis of elitism. It employs an unadorned and direct form of communication relating the product to the user. It is a vernacular "popular" approach, existing in the present tense, where its impact is most instantaneous and seductive. Its iconography and message are united in a universal language that speaks to us without camouflaging our guilty pleasures. It's about what we eat, what we use, and what we like.

JDA INC. RETAIL READY DESIGN

1 Product: earBudeez Earbud Packaging, Client: Audiovox Accessories Corporation, Distribution: USA, 2009

Bodie, Jay D., Skull Rojo, and Zoie Jane comprise the cast of the ear-Budeez collection. The colorful, silk-screened plastic clamshell packaging draws attention to the product as the earbuds double as the eyes of the youth culture-inspired characters.

MUCCA

2 Product: La Condesa Matches, Client: The Icon Group, Distribution: USA, 2007

Historical and contemporary cultural influences that are present in the vibrant Mexico City neighborhood of La Condesa are brought together by this packaging. Its typography reflects the tradition of hand-stenciled signage while a masked lucha libre wrestler meets those who open the matchbook with a grimace.

NEOSBRAND

3 Product: Obamitas, self-initiated, Distribution: Europe, 2009

STIR

4 Product: Middle Brau, Client: Middlebrow Comics, Distribution: USA, 2010, Material: color printouts on heavy paper stock

A clever play on words and an iconic logo were at the heart of the guerrilla promotional campaign for Middlebrow Comics. False fronts were placed on existing six-pack beer packaging in liquor store refrigerators.

IGOR MANASTERIOTTI & MIA MARIC

5 Product: Beer of Osijek Winter Edition, Client: Osjecka Pivovara (Osijek Brewery), Distribution: Croatia, 2010

Changing the color of the front label from classic gold to Christmas red, along with the small intervention to the bottleneck label, brought holiday flair to the winter edition of this Croatian beer.

1

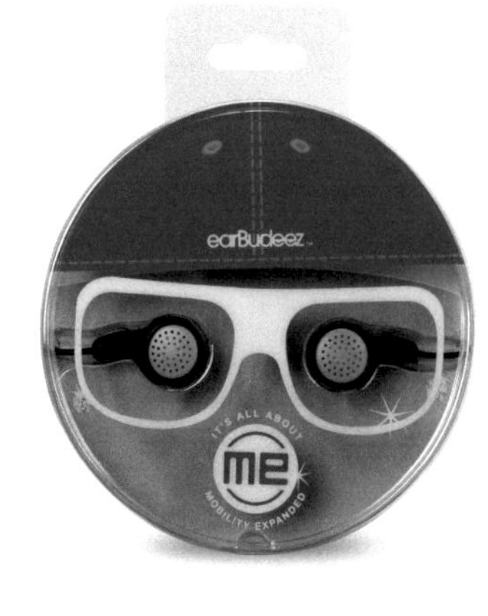

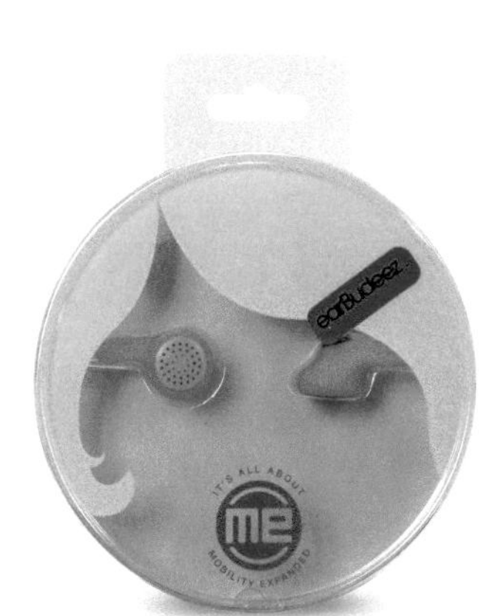

2

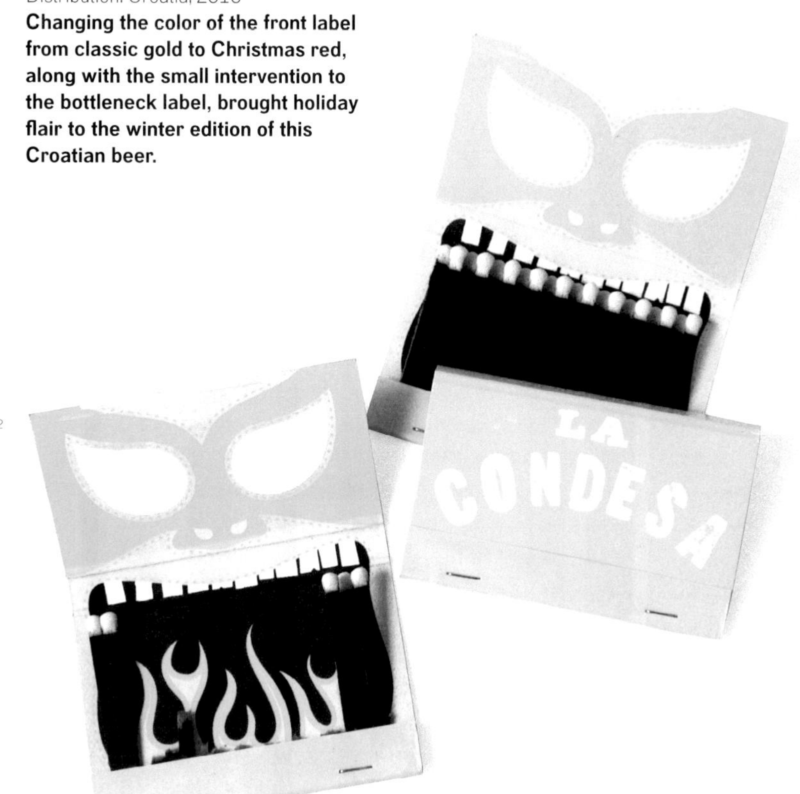

3

4

MIDDLE BRÄU
GREAT TASTE. LESS SHILLING.
MIDDLEBROW-COMICS.COM

MIDDLE BRÄU
MIDDLE BRÄU
Hoegaarden
$8.99
$8.99
$8.99
$8.99

5

OSIJEČKO

MOXIE SOZO

1. Product: Snackle Mouth, Client: Big Mouth Snack Co., Distribution: USA, 2010
2. Product: Bar Soap, Client: LEAP Organics, Distribution: USA, 2009
3. Product: Matcha Green Tea, Client: Jade Monk Beverage Co., Distribution: USA, 2010

1

2

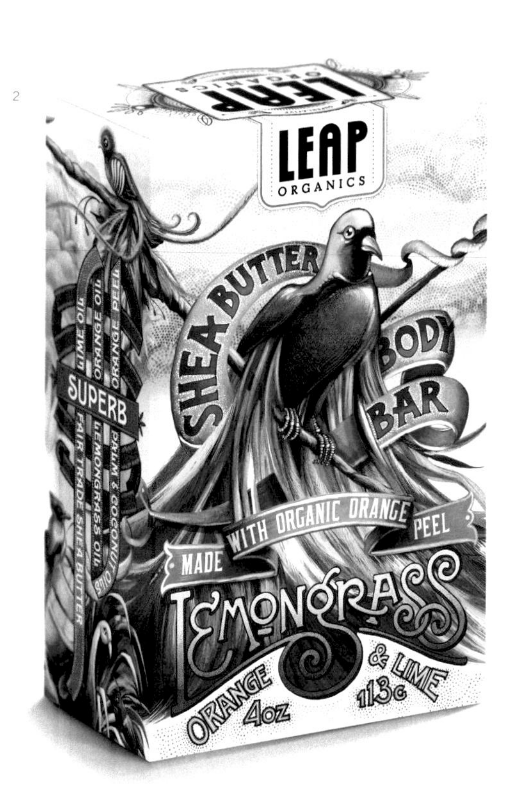

3

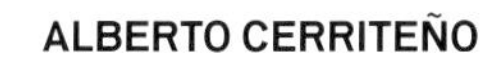

4

ALBERTO CERRITEÑO

4 Product: INQ Mini Cellphone, Client: INQ, Distribution: Europe, 2009

NTGJ

5 Product: The Olive Oil Experience Limited Edition, Client: Think Global Taste Local, Distribution: worldwide, 2011, Material: aluminum can and cardboard

The limited edition gift set of three types of olive oil from three different countries was conceived as an exceptional gift for those interested in the appreciation of fine food, culture, and art. Three young artists from each of the countries represented were invited to illustrate the collectible oil cans, depicting something that they love about their respective countries.

5

1

Mango
Pine apple
BEAR
baked fruit nibbles
under 100 cals
100% fruit

BEAR

2

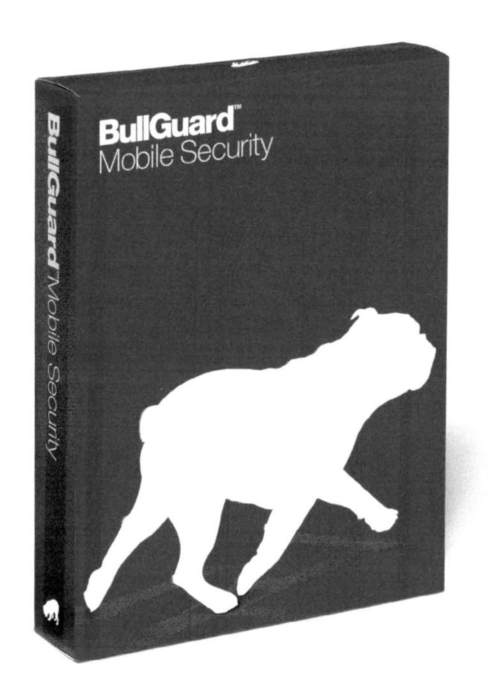
BullGuard™
Mobile Security
BullGuard™ Mobile Security

BullGuard™
Backup Security
BullGuard™ Backup Security

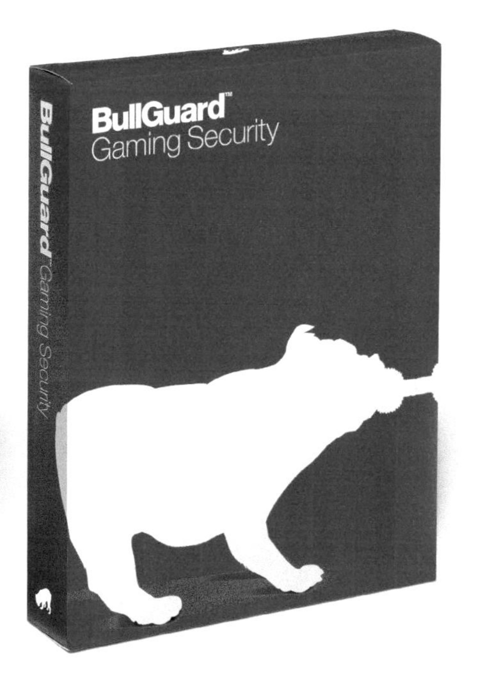
BullGuard™
Gaming Security
BullGuard™ Gaming Security

3

life
ALL ROUND BODY SCRUB
200 ML 6.8 FL.OZ e
BALANCING BODY LOTION
200 ML 6.8 FL.OZ e
PARTY FEET FOOT CREAM
150 ML 5.1 FL.OZ e
SUPER SMOOTH SHAVE CREAM
150 ML 5.1 FL.OZ e
MIGHTY MITTS HAND CREAM
PICTURE PERFECT FACIAL MOISTURISER
100 ML 3.3 FL.OZ e
WORK IT APD
UPLIFTING BODY LOTION
200 ML 6.8 FL.OZ e
JUMP START BATH & SHOWER GEL
350 ML 11.9 FL.OZ e

4

5

6

B&B STUDIO

1 Product: Bear Fruit Nibbles, Client: Bear, Distribution: UK, 2010

PURPOSE

2 Product: Bullguard packaging, Client: Bullguard, Distribution: worldwide, 2010, Material: 485 mic Incada silk

BOB DESIGN LTD

3 Product: Life NK , Client: SPACE.NK, Distribution: UK, 2009

A talisman belonging to Space NK founder Nicky Kinnaird now finds its way onto Life NK packaging, adding an unspoken personal touch to the cosmetic range. The bear is depicted in actions that reflect the name and mood of the particular product, while various colors reflect the different fragrance families.

CHRIS VON SZOMBATHY

4 Product: Flav'R Full, Client: Literary/Visual Artwork, Distribution: Canada, 2009, Material: cardstock, acrylic

SHU HUNG & JOE MAGLIARO

5 Client: Kaiserhonig, Distribution: Germany, 2010

MICHAEL GOLAN

6 Product: Anima 50 ml Bottle Labels, Client: Anima, Distribution: Israel, 2010-2011

An elegantly eye-catching as well as cost-efficient solution for the Anima organic animal care product line. The bottle labels can be produced using spot color printing with black and one pantone color, or with only three colors in CMYK, thus reducing printing costs even more.

MATS OTTDAL

1 Product: Birdy Juice, 2010
A colorful aviary was designed to appeal to young juice consumers. The inspiration for the birds' wings and feet derives from a playful twist to the collapsed shape of a standard juice carton.

1

2

IDENTITYWORKS
Henrik Hallberg
2 Product: Kaviar, Client: ICA, Distribution: Sweden, 2000, Material: metal standard tube

HELEN MARIA BÄCKSTRÖM
3 Product: NooDel, Student Project, Distribution: Sweden, 2011, Material: cardboard
NooDel integrates practical handling with an attractive motif that reflects its contents: a geisha with chopstick hairpins doubles as an easy-to-carry, microwavable container for instant Asian noodle soup.

3

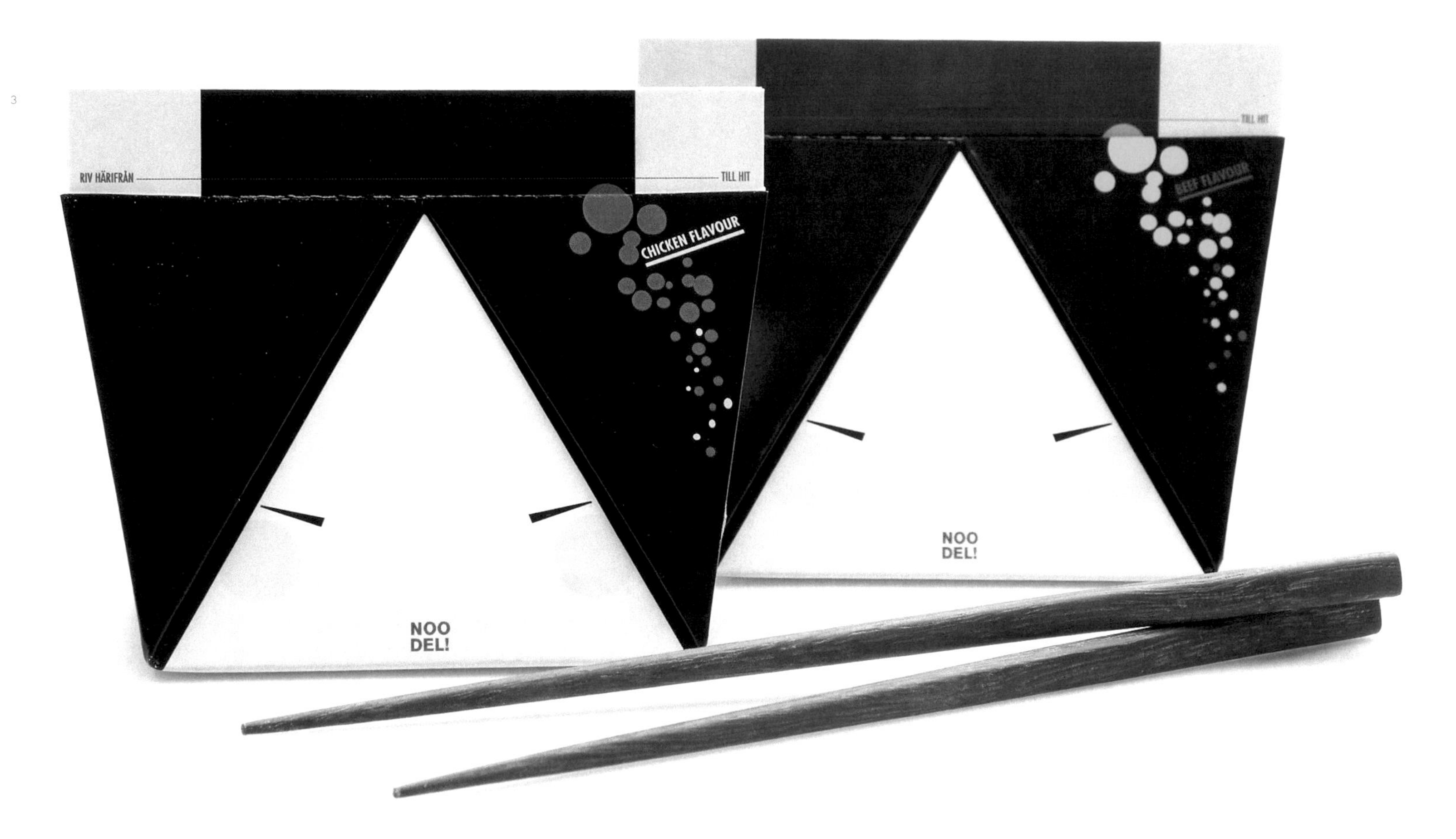

COARSE

Mark Landwehr, Sven Waschk

1 Product: Jaws, Client: coarse, Distribution: worldwide, 2009, Material: silkscreen printed wooden box, closed with woven elastic band, sponge inlay

ESRA OGUZ

2 Product: Kido Milk, Client: Kido, 2011

STUDIO KLUIF

3 Client: Hema, Distribution: The Netherlands, Belgium, Germany, 2010

1

2

3

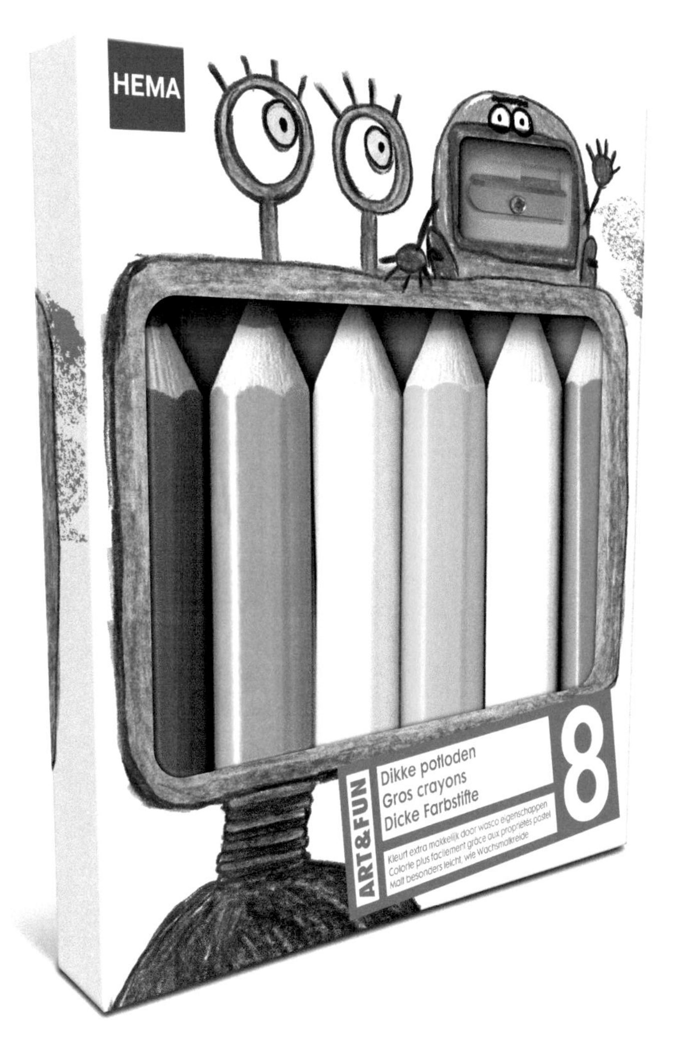

4

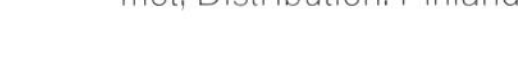

HASAN & PARTNERS

4 Product: Fazer Vilpuri, Client: Fazer Leipomot, Distribution: Finland, 2009

LOWE BRINDFORS

5 Product: NK Kids, Client: NK, Distribution: Sweden, 2010

Light-hearted and playful, this animal-inspired gift packaging for the inauguration of a new children's clothing section at a Swedish department store was designed to profile the department as the number one destination for kids.

5

AH&OH STUDIO

1 Product: Scent Stories, self-initiated, 2010
Drawing inspiration from masterpieces of literature, and their dark and distinctive male characters, the range of men's perfume named after famous writers was filled into white bottles resembling a cross between vintage perfume bottles, and the classic inkwell. Marked by bold black lettering, each bottle is topped by the head of a character derived from one of the featured author's great works.

WILLIAMS MURRAY HAMM

2 Product: Comic Relief Indian Sauces, Client: Comic Relief, Distribution: UK, 2011

STUDIO KLUIF

3 Product: Bolletje Wol Packaging, Client: Bolletje Wol, Distribution: The Netherlands, 2008
The packaging for this series of handmade children's hats reflects its contents in more ways than one. One side presents a portrait of a child wearing a hat whose shape is represented by the form of the box. The other side of the package depicts the mountains of Nepal, where the hats are produced.

4 Product: Office Supplies Packaging, Client: Hema, Distribution: The Netherlands, Belgium, Germany, 2002
An amusing, eye-catching, and unmistakably familiar motif is used here to make the in-house office paper of this Dutch department store chain stand out above the rest.

1

2

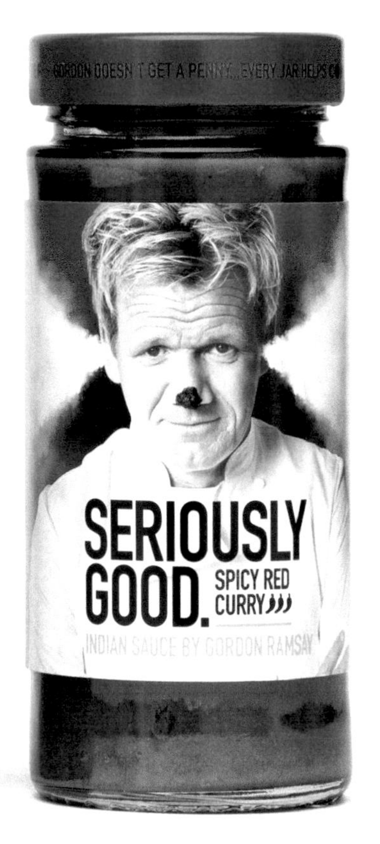

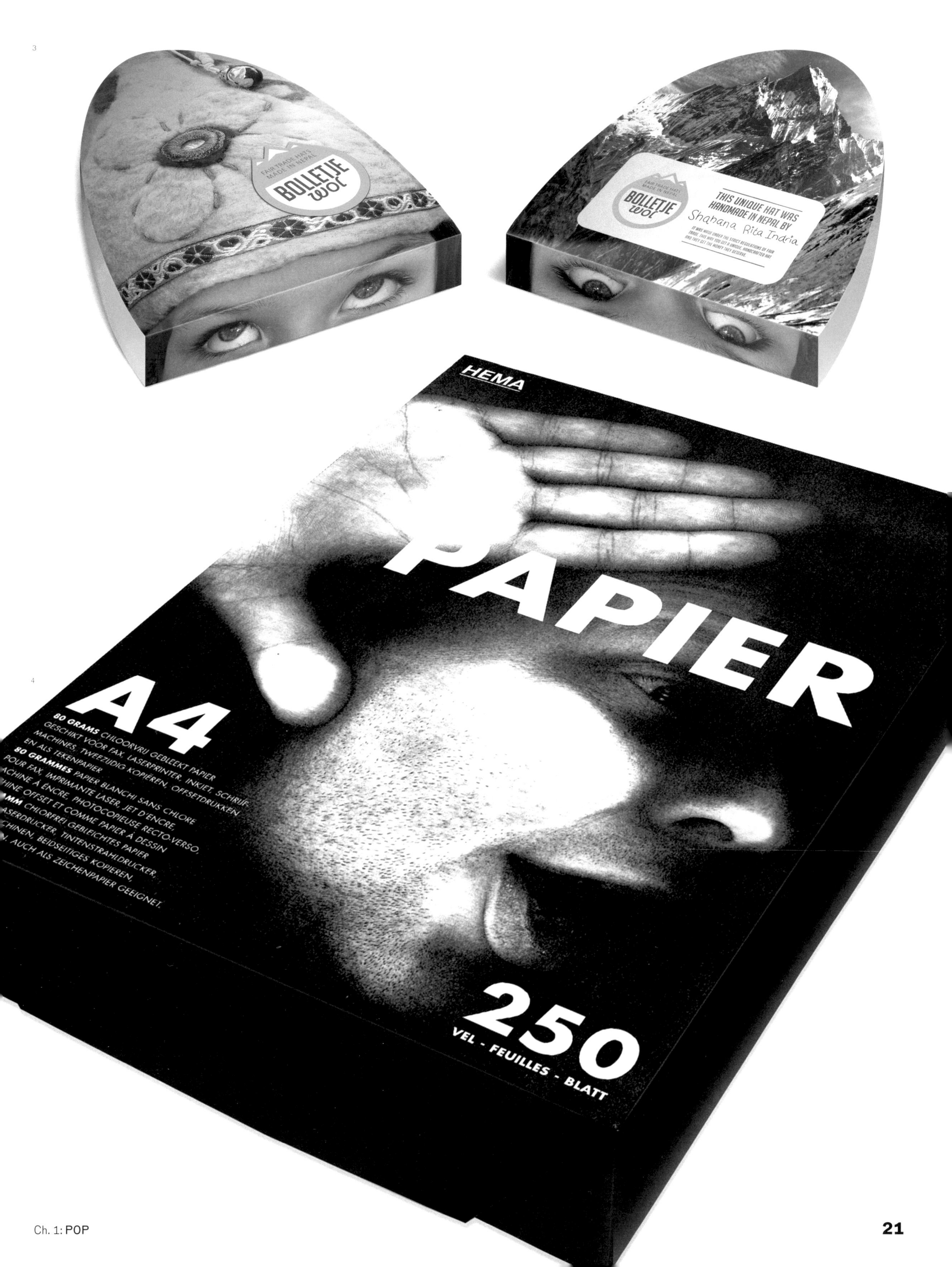
3
FAIR TRADE HATS MADE IN NEPAL
BOLLETJE wol
FAIR TRADE HATS MADE IN NEPAL
BOLLETJE wol
THIS UNIQUE HAT WAS HANDMADE IN NEPAL BY
Shabana Rita Indria
HEMA
PAPIER
4
A4
80 GRAMS CHLOORVRIJ GEBLEEKT PAPIER
GESCHIKT VOOR FAX, LASERPRINTER, INKJET, SCHRIJF-
MACHINES, TWEEZIJDIG KOPIËREN, OFFSETDRUKKEN
EN ALS TEKENPAPIER
80 GRAMMES PAPIER BLANCHI SANS CHLORE
POUR FAX, IMPRIMANTE LASER, JET D'ENCRE,
ACHINE À ENCRE, PHOTOCOPIEUSE RECTO-VERSO,
HINE OFFSET ET COMME PAPIER À DESSIN
MM CHLORFREI GEBLEICHTES PAPIER
ASERDRUCKER, TINTENSTRAHLDRUCKER,
CHINEN, BEIDSEITIGES KOPIEREN,
, AUCH ALS ZEICHENPAPIER GEEIGNET.
250
VEL - FEUILLES - BLATT

EVERYTHING

1 self-initiated, Distribution: New Zealand, 2010, Material: wooden crate

A jovial Christmas narrative on the subject of "everything in moderation" adorns the bottles of this boxed beer set. While the text on the front labels soberly present the moral of the story and product information, the back labels feature a comical visual narrative beginning with Santa's Christmas night, and ending in an utter mess.

ROXY TORRES

2 Product: Coyote Fireworks, Student Project, 2010, Material: wood, canvas, newsprint paper, and illustration board

This student packaging project developed for a fictional line of fireworks takes its inspiration from a classic miner look and feel–using burlap sacks, vernacular type traditionally used in handpainted signage, and a branded wooden box in handcrafted, vintage style for transportation.

COYOTE
FIREWORKS

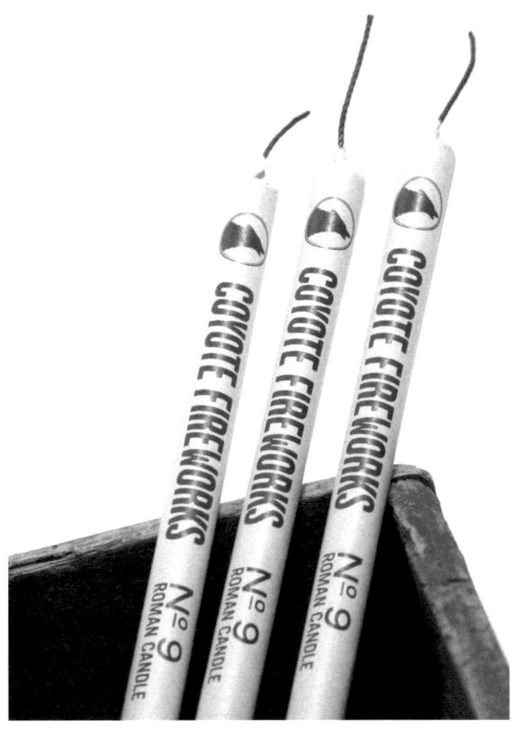
COYOTE FIREWORKS
Nº 9
ROMAN CANDLE
COYOTE FIREWORKS
Nº 9
ROMAN CANDLE
COYOTE FIREWORKS
Nº 9
ROMAN CANDLE

COYOTE
Nº 35
PREMIUM
SHELLS
COYOTE FIREWORKS

COYOTE
FIREWORKS
3 UNITS
Nº 35
PREMIUM
SHELLS
WARNING SHOOTS FLAMING BALLS
COYOTE FIREWORKS
COYOTE FIREWORKS
COYOTE
FIREWORKS
Nº 4
COLORED
SPARKLERS
Nº 2
SMOKE
BOMBS

FRAN ALDEA

1 Product: Las Marías de La Cumbre, Client: UDD, Distribution: Chile, 2010, Material: cardboard

THE STROHLS

2 Product: Bunny Bar Brand Identity & Packaging, Client: 18 Rabbits, Distribution: USA, 2010, Material: paperboard

FREDDY TAYLOR

3 Product: Bean Bags, Client: Geo Organics, Distribution: Britain (proposed), 2011, Material: translucent plastic, heat sealed, resealable, recyclable and microwave safe

ABLE

4 Product: One Village Coffee, Client: One Village Coffee, Distribution: USA, 2010
The coffee bag developed for One Village Coffee aims to package village-inspired sentiments into a customer experience. Texts and illustrations provide information on the eco-minded company while emphasizing the importance of community in a light-spirited and personal tone. As the budget allowed for only a single bag, customized labels were designed to identify each type of coffee.

MIKA KAÑIVE

5 School Project, Distribution: Spain, 2010, Material: PET 100%

1

2

3

4

ONE VILLAGE
COFFEE
Specialty Roaster
ONEVILLAGECOFFEE.COM
ARTIST BLEND
BOLD • EARTHY • FRUITY
LOCALLY ROASTED
WE CALL IT A VILLAGE.
WELCOME & ENJOY.
1. THE BEST BEANS
We start by choosing beans from the top 5% of the harvest (top of their class you might say)
Next we find the best roast for each type of bean. Call us romantics, but we believe every bean has a perfect roast to match.
3. ALL THE PEOPLE
It's our belief that good beans start and end with good people, the growers, pickers, roaster, and the person who put this bag on the shelf.
ONE VILLAGE COFFEE CO.
Certified
B

5

Limón
& Nada
Bebida natural
compuesta al 100%
por zumo de
frutas; con
vitaminas y
minerales.
1L

Naranja
& Nada
Bebida natural
compuesta al 100% por zumo
de frutas; con vitaminas y
minerales.
1L

WALDO PANCAKE (JIM SMITH)

Client: Puccino's, Distribution: UK, 2009

A unique design of cups, napkins, sugar packets, bags, signs, and more has shaped and helped to establish the identity of this coffee shop. The personal, hand-drawn style, and the predominance of words over images consistently capture the designer's quirky and irreverent sense of humor.

Put grain under mattress of enemy.
Puccino's
white sugar for commoners
Rubbish beanbag.
As seen on The Larry Puccino Show.
Global Warming is all my fault.
Google me.
Not one of those handwarmers.
Not you again.
So this is how it ends.
Tear.
Rattle in background for atmos.
Pour coffee into sachet.

Puccino's
brown sugar for snobs
1980's style sachet.
Fold corner over and continue reading later.
Open using charm/gentle persuasion.
Think of better thing to write.
This is not a white sugar sachet.
Complain about printing error
Do montage of good times together in head after opening.
Spend ages reading then wonder if was waste of life or not.
Squint to read.
A4 sachets coming soon.
Take photo of and post on internet.

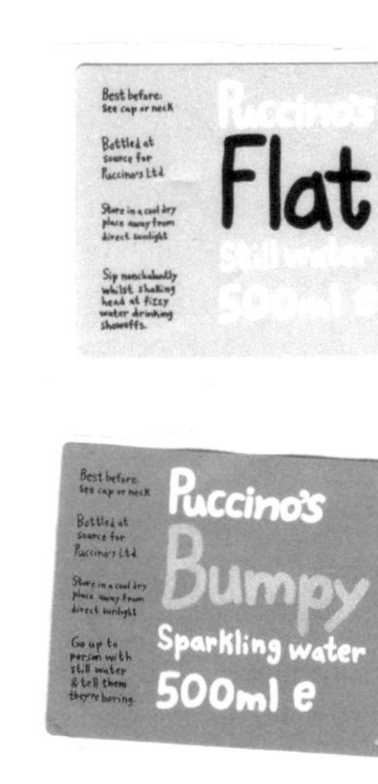

Best before: See cap or neck
Bottled at source for Puccino's Ltd
Store in a cool dry place away from direct sunlight
Puccino's
Flat
Still water
500ml e

Best before: See cap or neck
Bottled at source for Puccino's Ltd
Store in a cool dry place away from direct sunlight
Go up to person with still water & tell them they're boring
Puccino's
Bumpy
Sparkling water
500ml e

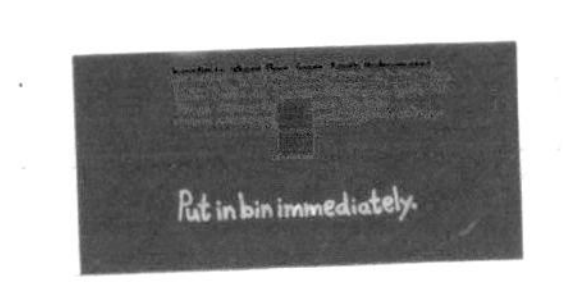

Puccino's
Stupid little biscuit

Put in bin immediately.

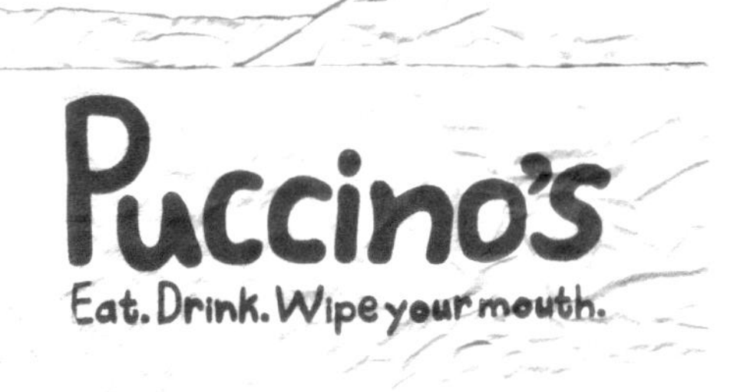

Puccino's
Eat. Drink. Wipe your mouth.

New napkin design
Our designer sketches all his ideas on the backs of napkins.
Except for when he's designing napkins.
'I wouldn't want to risk the irony' he says.

Coffee is SO over.

any 4 tapas, warm pitta & bottle of house wine £15.95 mon to fri
Puccino's
Blame self for non flowingness of chat at table.

Crumbs

Confuse waiter for long lost sibling.

BOB HELSINKI

1 Product: KC Professional–Four Reasons Take Away Color, Client: Miraculos, Distribution: Finland, 2010

LAURA BERGLUND

2 Product: Rêve, Kansas City Art Institute student project, Distribution: USA, 2009, Material: kraft paper, hand silk-screening, 80 lb. text label system

The brand aesthetic developed pairs sophisticated materials with lighthearted typography and cheerful bursts of color to attract a new audience to the upscale hot air balloon ride company.

MARTIN POPPELWELL

3 Product: The People's Wine, Client: Constellation, Distribution: New Zealand and Canada, 2010

SARA NICELY

4 Product: Flour Pot Bakery Identity System, Student Project, 2010

SERGE RHÉAUME

5 Product: Nos Desserts d'Antan, student-project, 2011

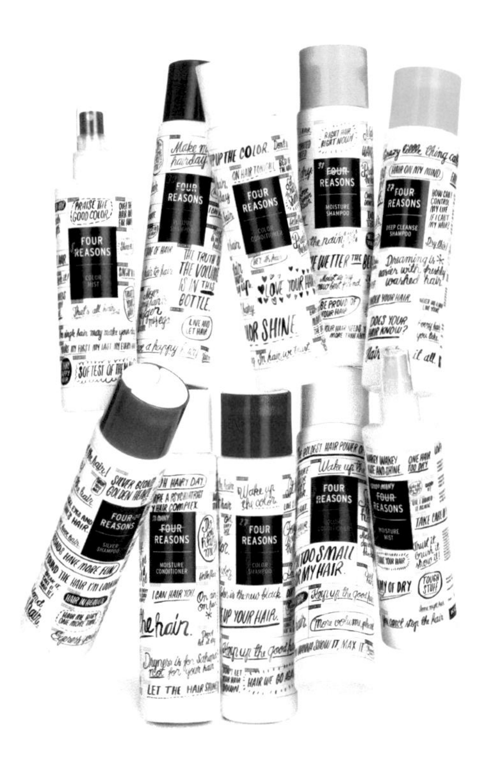

1

2

3

The People's
Sauvignon
Blanc
Pinot Gris
Crafted in
MARLBOROUGH
NewZealand 2010
The People's
Chardonnay
CRAFTED in
the Wairau
Valley
Marlborough
NewZealand 2009
The People's
Pinot Gris
Crafted at
Corner 50
Winery,
Hawkes Bay
NewZealand 2010
The People's
Sauvignon
Blanc
Crafted in
the Awatere
Marlborough
NewZealand 2010
The People's
Pinot Noir
Crafted in
ALEXANDRA
Central Otago
NewZealand 2009
The People's
Crafted at
Corner 50
Winery,
Hawkes Bay
NewZealand 2009

4

FLOUR POT
BREADS
PASTRIES
BAGUETTE
ROLLS
CAKES
DESSERTS
QUICHE
SANDWICHES
FLOUR POT
FLOUR POT LUNCH
QUICHE
CROISSANTS
SANDWICHES
DRINKS

5

POUDING
CHÔMEUR
24

GRAND-PÈRE
DANS L'SIROP
DES FRIANDISES AUX
SAVEURS DES DÉLICES
QUÉBÉCOIS
24

MICHA WEIDMANN STUDIO

1 Product: Gift and Mail Boxes, Client: Davina Peace, Distribution: UK, 2010

BASEDESIGN

2 Product: Graanmarkt 13

P576

3 Product: baguette paper bags, Client: paniqueso, Distribution: Colombia, 2010, Material: paper

STUDIO RAŠIĆ

4 self-initiated, Distribution: Croatia, 2010

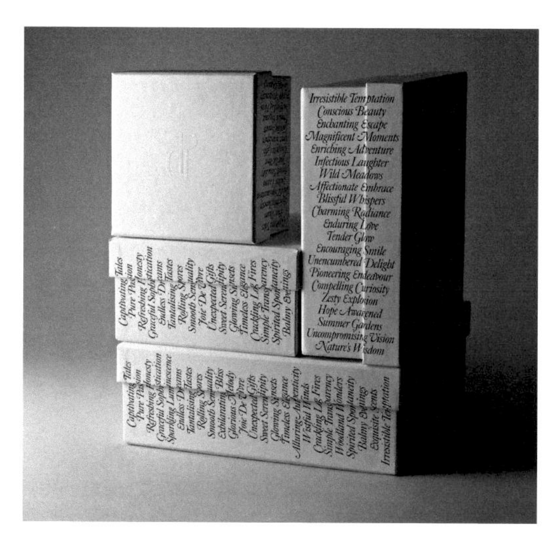

1

2

3

HAND MADE
IN STUDIO RAŠIĆ!

CHRIS VON SZOMBATHY

1 Thinking Poptimistically, self-initiated, Distribution: Canada, 2008–2010, Material: glass, plastic, acrylic

The range of hand-painted glass containers elevates the notion of bottle collecting to another level. A total of 77 bottles embellished with various words, and corresponding images allow for a plethora of playful phrases or sentences to express one's sentiments, and ideas.

ANDREI POPA

2 Product: Brisk, self-initiated, 2009

1

2

brisk

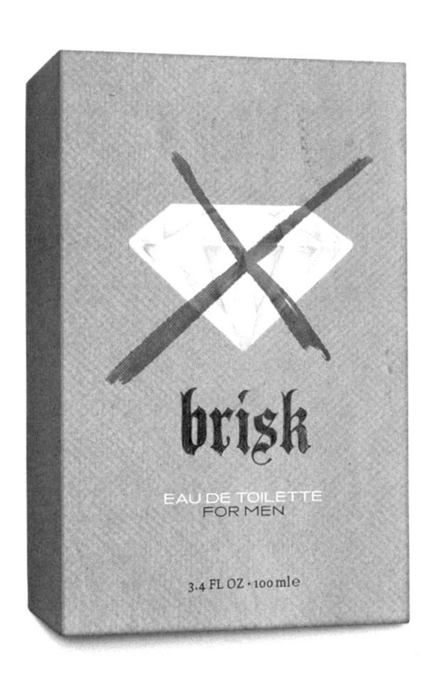
brisk
EAU DE TOILETTE
FOR MEN
3.4 FL OZ · 100 ml e

BODY & HAIR
WASH
FOR MEN
200ml · 6.6 fl.oz.

brisk
BODY & HAIR
WASH
FOR MEN
200ml · 6.6 fl.oz.

Þorsteinn
Þorsteinn
Þorsteinn
Þorsteinn
Þorsteinn

THORLEIFUR GUNNAR GÍSLASON, GEIR OLAFSSON, HLYNUR INGOLFSSON

Product: Thorsteinn Beer brand, self-initiated, 2010, Material: clear glass bottles

Creating a unique platform for graphic designers as well as challenging typical beer labelling tendencies was the goal of this packaging project. The single-color graphics establish a unifying element that can accommodate an annually changing range of ten designs per collection.

1

TOCOBAGA
RED ALE
ABV 7.2%
9.4 FL. OZ.

2

SINCE 2003
COUNTRY
CIDER
FOUR.FIVE PERCENT
330ml
STRAWBERRY
SCRUMPIN'

SINCE 2003
COUNTRY
CIDER
FOUR.FIVE PERCENT
330ml
APPLE

FINEST QUALITY
ORGANIC GOODNESS
SINCE 2003
COUNTRY
CIDER
FOUR.FIVE PERCENT
330ml
PEAR

3

FORÅRS
ALE
RIBE BRYGHUS
BLOND
ALE
RIBE BRYGHUS
BROWN
ALE
RIBE BRYGHUS
BLACK
ALE
RIBE BRYGHUS
JULE
ALE
RIBE BRYGHUS

4

HEFEWEIZEN
Weizen
·UP·
HELLES
Helles
·BELLES·
ALTBIER
Alt
·EGO·

5

6

ALEXANDER KUSIMOV

1 Product: Tocobaga Red Ale, Client: Cigar City Brewing Co., Distribution: USA, 2011

ILOVEDUST

2 Product: Country Cider, 2010

MADS JAKOB POULSEN

3 Product: Ribe Ale, Client: Ribe Micro Brewery, Distribution: Denmark, 2010, Material: label on uncoated off-white stock

D.STUDIO

4 Product: German Beers, Client: Westerham Brewery, Distribution: UK, 2011, Material: paper label

ILOVEDUST

5 Product: Sea Cider, 2010

CATHERINE BOURDON

6 Product: Sugar Skull, Distribution: Canada, 2010

ATIPUS

1 Product: Vi Novell, Client: Celler el Masroig, Distribution: Spain, 2010

SUPPLY

2 Product: Rochdale Cider, Client: McCashin's, Distribution: New Zealand, 2010

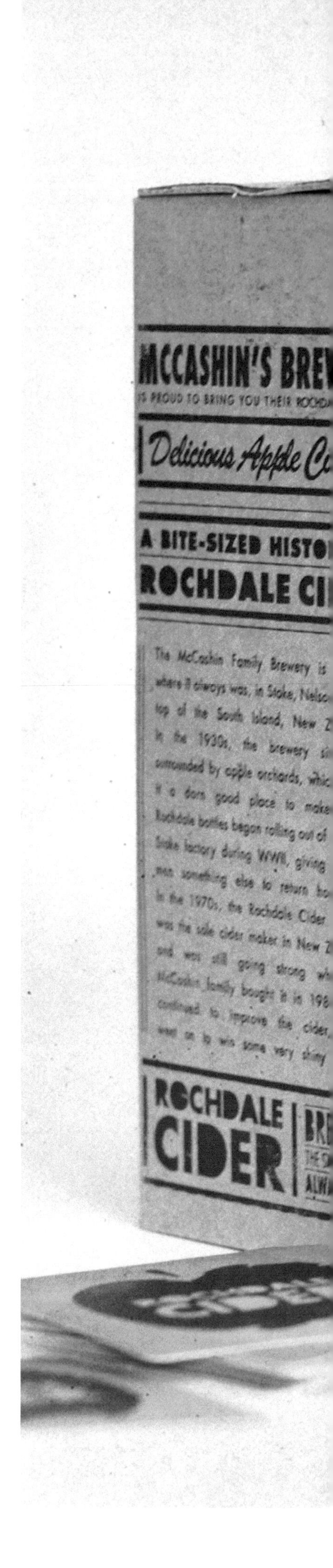

ROCHDALE
CIDER
TRADITIONAL
SPARKLING, MEDIUM-DRY
TRADITIONAL CIDER.
WELL-ROUNDED FLAVOUR
+ LIGHT, CRISP FINISH
LOVINGLY MADE IN NEW ZEALAND BY
MCCASHIN'S BREWERY
A TRADITION
SINCE 1941
ROCHDALE CIDERS ARE PROUDLY BREWED LIKE THEY USED TO BE. NOTHING BUT THE MOST DELISH TREE-RIPENED NZ GROWN APPLES &PURE ARTESIAN WATER
WWW.ROCHDALECIDER.CO.NZ
5.0% ALC VOL
1.3
4x330MLS
LIGHTLY EFFERVESCENT
PERFECT ANYTIME
GLUTEN FREE
100%
Thirst-Quenching
GUARANTEED!
NØ PRESERVATIVES WHATSOEVER
TRADITIONAL
5.0% ALC/VOL
NØ PRESERVATIVES
ROCHDALE
CIDER

PENTAGRAM DESIGN LIMITED
Harry Pearce
1 Product: Budgens/Londis Spirit Labels, Client: Musgrave Retail Partners GB, Distribution: UK, 2010, Material: paper on glass

MASH
2 Client: Alpha Box and Dice, Distribution: Australia, USA, UK , 2010
An alphabet of wine was created for this young wine producer. For each of the 26 letters, 26 stories and 26 pieces of label art were developed to reflect the eclectic ideas and personality of the owners.

PARALLAX DESIGN
3 Product: The Scarlet Letter, Client: Henry's Drive Vignerons, Distribution: Australia, USA, Canada, 2010, Material: uncoated textured stock, high build screen, 750 ml

1

2

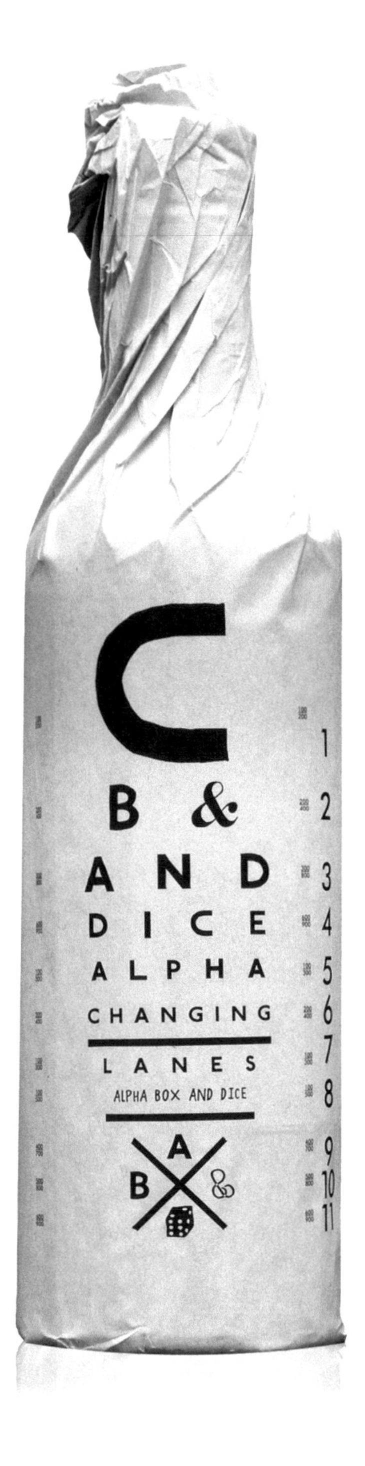

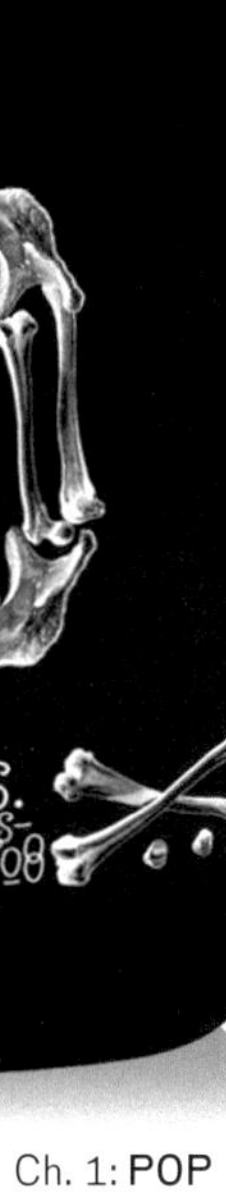

ENIGMA
2007 BARBERA/CABERNET SAUVIGNON
750ML

3

The
SCARLET
LETTER
Sparkling Shiraz

BUDDY

1 Product: Mulled Wine, Client: Buddy, Distribution: UK, 2009

"The more you drink, the merrier the message" was the idea behind these Christmas mulled wine bottle produced for clients and friends. Printed directly onto the wine bottle was a holiday greeting as well as horizontal lines suggestive of typical measuring jug graphics.

CHRIS TRIVIZAS

2 Product: Fygein Adynaton, Client: Chris Trivizas, Distribution: Greece, 2009

STUDIO RAŠIĆ

3 Product: Medicine wine, self-initiated, Distribution: Croatia, 2010

Designed as a Christmas gift for clients from the pharmaceutical industry, Medicine Wine features labels reminiscent of Croatian medication packaging.

PARALLAX DESIGN

4 Product: Morse Code, Client: Henry's Drive Vignerons, Distribution: Australia, USA, Canada, 2009, Material: coated stock, matt varnish with high build screen

Following the vineyard's general postal theme branding, this range of wines pays homage to the first electronic mail system by printing the wine varietals on the label in Morse code.

DORIAN

5 Product: El Buscador & El Guía, Client: Finca de la Rica, Distribution: Spain, 2011

1

2

3

MALVAZIJA 12,9
MALVAZIJA
Malvazija
kvalitetno vino
0,75 l
Malvazija
MALVAZIJA 12,9

4

Morse Code
Padthaway
Chardonnay
2009
Morse Code
Padthaway
Shiraz
2008

5

El Buscador
El Guía

BENDITA GLORIA
Product: Casa Mariol Wine Collection, Client: Casa Mariol, Distribution: Spain, Poland, USA, 2010, Material: coated offset printed labels
Reflecting the client's philosophy of producing natural, quality wines without the pretension to luxury often seen in the wine industry, these wine bottle labels were designed with a no-frills, homemade aesthetic that makes use of everyday desktop publishing tools such as MS WordArt, Excel, and ClipArt. The wines are simply identified on the label according to their grape variety and age, eschewing romantic naming conventions.

CABERNET SAUVIGNON
CRIANÇA
14 MESOS
CASA MARIOL
SAMSÓ
CRIANÇA
16 MESOS
CASA MARIOL
MERLOT
RESERVA
24 MESOS
CASA MARIOL
CABERNET SAU-VIG-NON
RESERVA
26 MESOS
CASA MARIOL
SYRAH
RESERVA
22 MESOS
CASA MARIOL

GARNA-TXA BLANCA
BLANC
JOVE
CASA MARIOL

GARNA-TXA NEGRA
ROSAT
JOVE
CASA MARIOL

GARNA-TXA NEGRA
NEGRE
JOVE
CASA MARIOL

B&B STUDIO

1 Product: Urban Fruit, Client: Urban Fruit, Distribution: UK, 2010

DESIGNERS UNITED

2 Product: Tea Route Iced Tea, Client: The Tea Route, Distribution: Greece, 2008

TURNSTYLE

3 Product: DRY Soda Bottles, Client: DRY Soda, Distribution: USA, Canada, 2010, Material: clear labels on glass bottles

TRULY DEEPLY

4 Client: Gelati Sky, Distribution: Australia, 2010

Combining Italian icons with kaleidoscopic images of the featured flavors in a surrealistic montage, the packaging of this boutique gelato company communicates its slogan: "It's what dreams taste like."

1

2

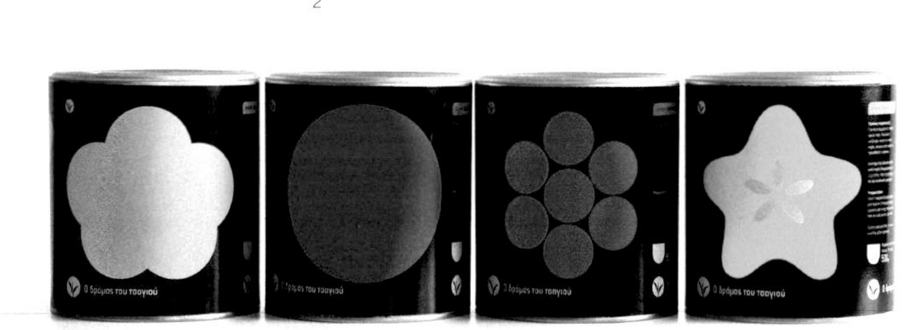

3

4

MON
SORBET

PASSION
FRUIT
SORBET
500mL

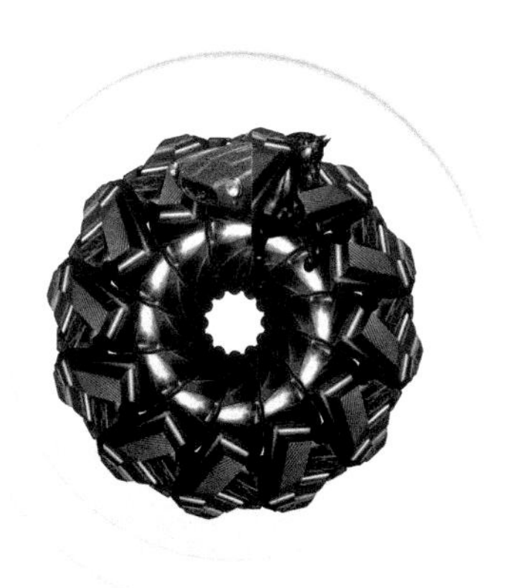

PEANUT
UTTER DELIGHT
GELATO
500mL

HOColate
GELATO
500mL

HONEY-
COMB
CRUNCH
GELATO
500mL

KIMBERLY-CLARK DESIGN
with Hiroko Sanders
Client: Kimberly-Clark Corporation/
Kleenex, Distribution: USA, 2011
For the Kleenex summer packaging campaign, the traditional cube-shaped facial tissue carton was replaced with a triangular wedge shape, and an assortment of food motif prints embodying the spirit of summer. The watermelon carton was the first to sell out.

Kleenex
Kleenex
Kleenex
Kleenex
Kleenex
70 2-PLY WHITE TISSUES
Kleenex

LINDSEY FAYE SHERMAN

1 Product: Back to Basics, Client: Watch & Grow Food Co., Distribution: Australia, 2010, Material: plastic pouch and printed label

JUST BE NICE STUDIO

2 Product: Premium Carrot, Client: Vegetoria, Distribution: Russia, 2010

PROMPT:/DESIGN

3 Product: Here! Sod, Client: Here! Sod, Distribution: Thailand, 2010, Material: foam, plastic, paper, fabric

2

1

เฮียสด
Here! Sod
Baguette
วันผลิต
ควรบริโภคก่อน
ขนาดสุทธิ
15.05.2010
15.05.2010
P1
ราคา
รวม
(บาท)
เฮียสด งดเชื่อ เบื่อทวง
เฮียสด
Here! Sod
resh Water
เฮียสด
Here! Sod
Beef
วันผลิต
ควรบริโภคก่อน
ขนาดสุทธิ
31.03.2010
31.03.2010
ราคา
รวม
(บาท)
เฮียสด งดเชื่อ เบื่อทวง
เฮียสด
Here! Sod
Chinese Cabbage
วันผลิต
ควรบริโภคก่อน
ขนาดสุทธิ
31.03.2010
31.03.2010
XL
ราคา
รวม
(บาท)
เฮียสด งดเชื่อ เบื่อทวง

SCHNEITER MEIER AG

1 Product: Migros Sélection, Client: Migros, Distribution: Switzerland, 2005

PETER SCHMIDT GROUP GMBH

2 Product: REWE Feine Welt, Client: REWE Markt GmbH, Distribution: Germany, 2009

1

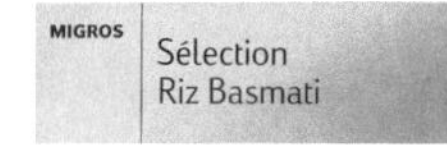

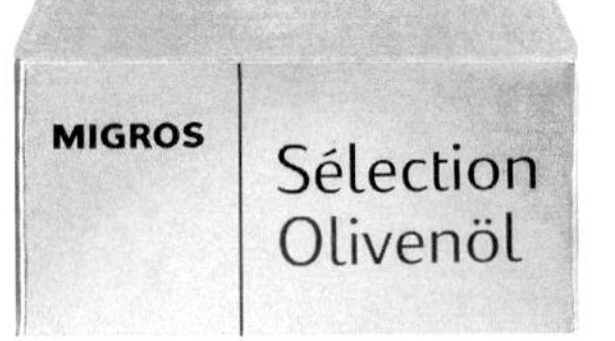

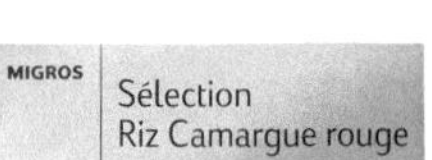

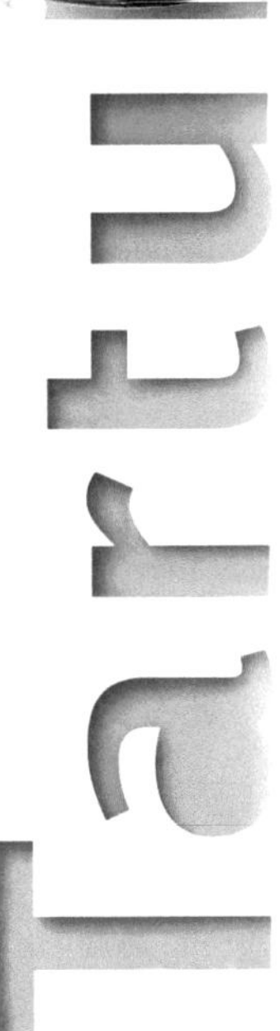

2

MIGROS
Sélection
Assam
Golden

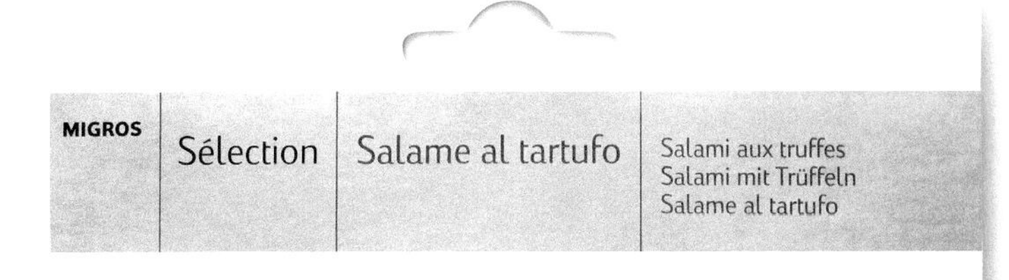

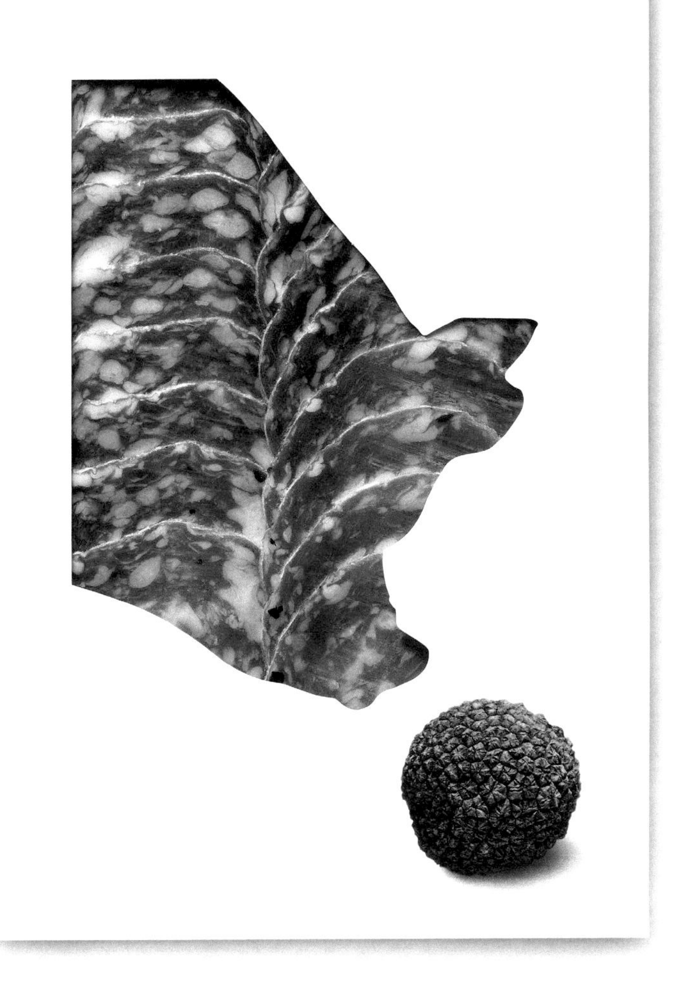

AT PACE

1 Product: Moya, Client: Moya, Distribution: South Africa, 2010

DESIGNERS UNITED

2 Product: Tea Route Tea Pouches, Client: The Tea Route, Distribution: Greece, 2010

1

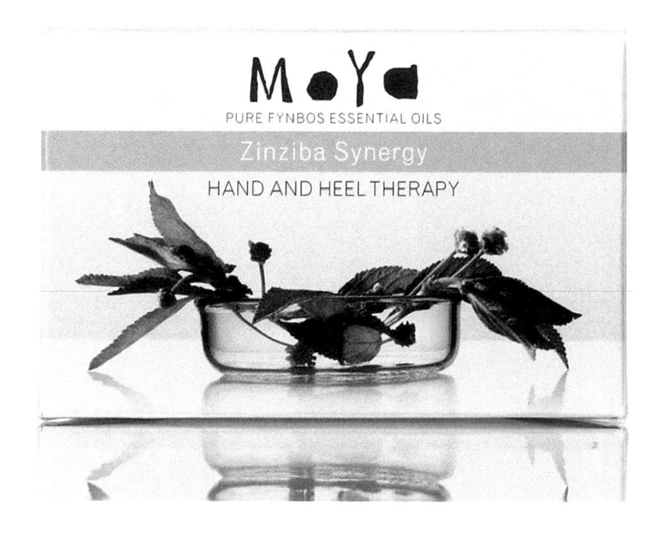

2

3

BLOOM DESIGN AGENCY

3 Product: Knackered Cow, Grumpy Cow, Lazy Cow, Horny Cow, Wild Cow, Moody Cow, Client: Cowshed, Distribution: Europe, Asia, Middle East, USA, 2007

SEK DESIGN

Client: AMS/Euro Shopper, 2010

1 Before

2 After

A no-nonsense, contemporary, and consistent packaging concept was devised for the overhaul of this European discount supermarket brand. Care was taken to select an easy-to-recognize design that would suit different cultural environments, and various printing techniques.

1

2

VANILLA
ICE CREAM
VANILJAJÄÄTELÖ · VANILJGLASS
1 l / 475 g

BROWN
SAUCE x3

GREEN
PEPPER
SAUCE x3

BEARNAISE
SAUCE x3

NATURAL
CRISPS
PAPRIKA
CRISPS
NATURAL
CRINKLE CUT
CRISPS
SOUR CREAM
& ONION
CRINKLE CUT
CRISPS

STRAWBERRY
DILUTABLE
FRUIT DRINK
BLACKCURRANT
DILUTABLE
FRUIT DRINK
GRENADINE
DILUTABLE
FRUIT DRINK
ORANGE
DILUTABLE
FRUIT DRINK

EXTRA
STRAWBERRY JAM
450 g
EXTRA
APRICOT JAM
450 g
EXTRA
RASPBERRY JAM
450 g
ORANGE MARMELADE
450 g
EXTRA
CHERRY JAM
450 g

1

MARINDE VAN LEEUWEN-FONTEIN

1 Product: Dutch Design Cheese-Lace,
Client: DesignKaas/Dutch Design Cheese,
Distribution: Europe, 2009

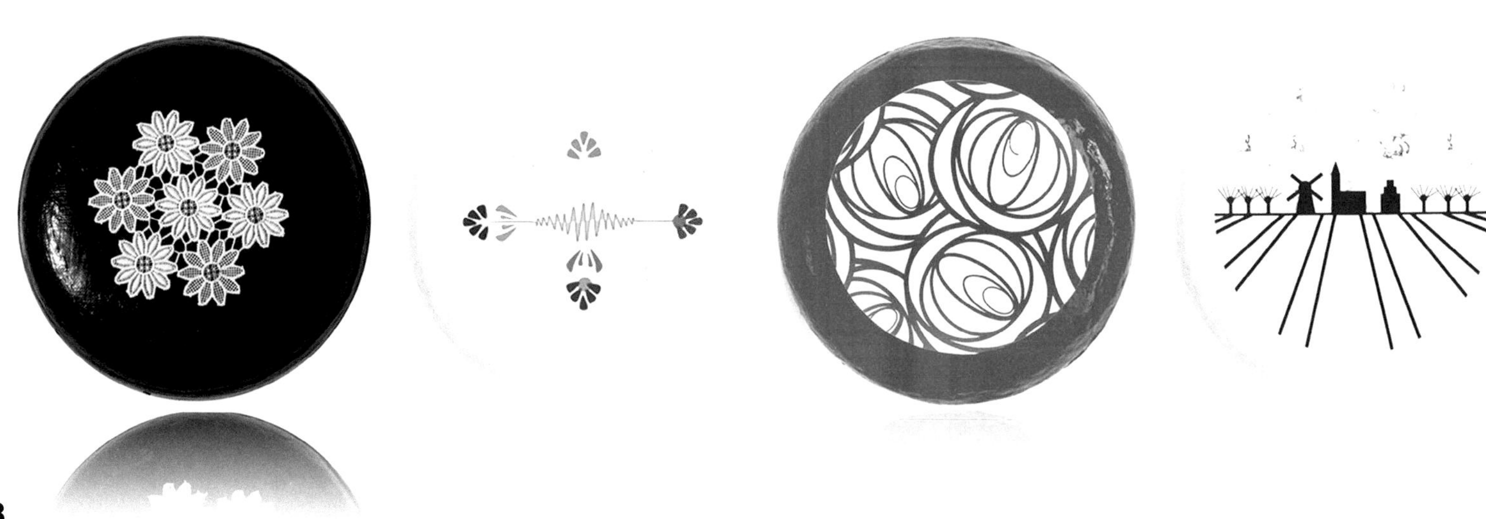

THE METRIC SYSTEM DESIGNSTUDIO, OSLO

2 Client: Illegal Burger, Distribution: Norway, 2011

MIND DESIGN

3 Client: What On Earth, Distribution: UK, 2009, Material: printed card boxes with embossing

The packaging concept for this organic food producer includes the combination of numerous hand-made, linocut illustrations on a plain background. The establishment of an image database facilitates the design of future products in the range.

2

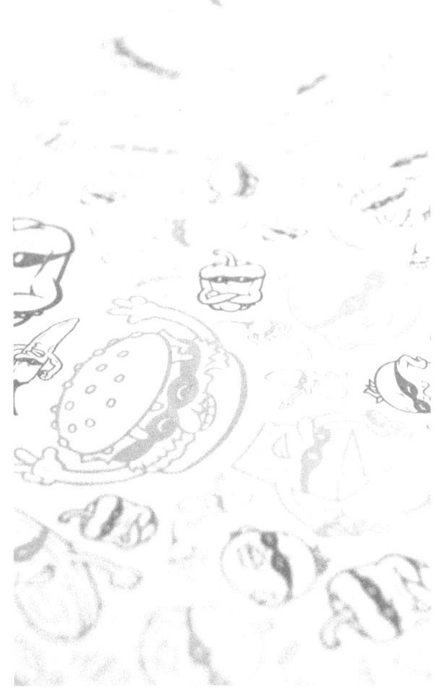

3

JONES KNOWLES RITCHIE

1 Product: Porkinson Redesign, Client: Kerry Foods/Porkinson, Distribution: UK, 2010

THE PARTNERS

2 Product: Mr Singh's Bangras, Client: Daljit Singh, Distribution: UK, 2009

Instead of fancy packaging, the product itself is branded, literally. Each of the gourmet Indian sausages is printed with a unique mendhi henna pattern in edible ink. The sausages are showcased in a tray covered by a simple protective sleeve.

2

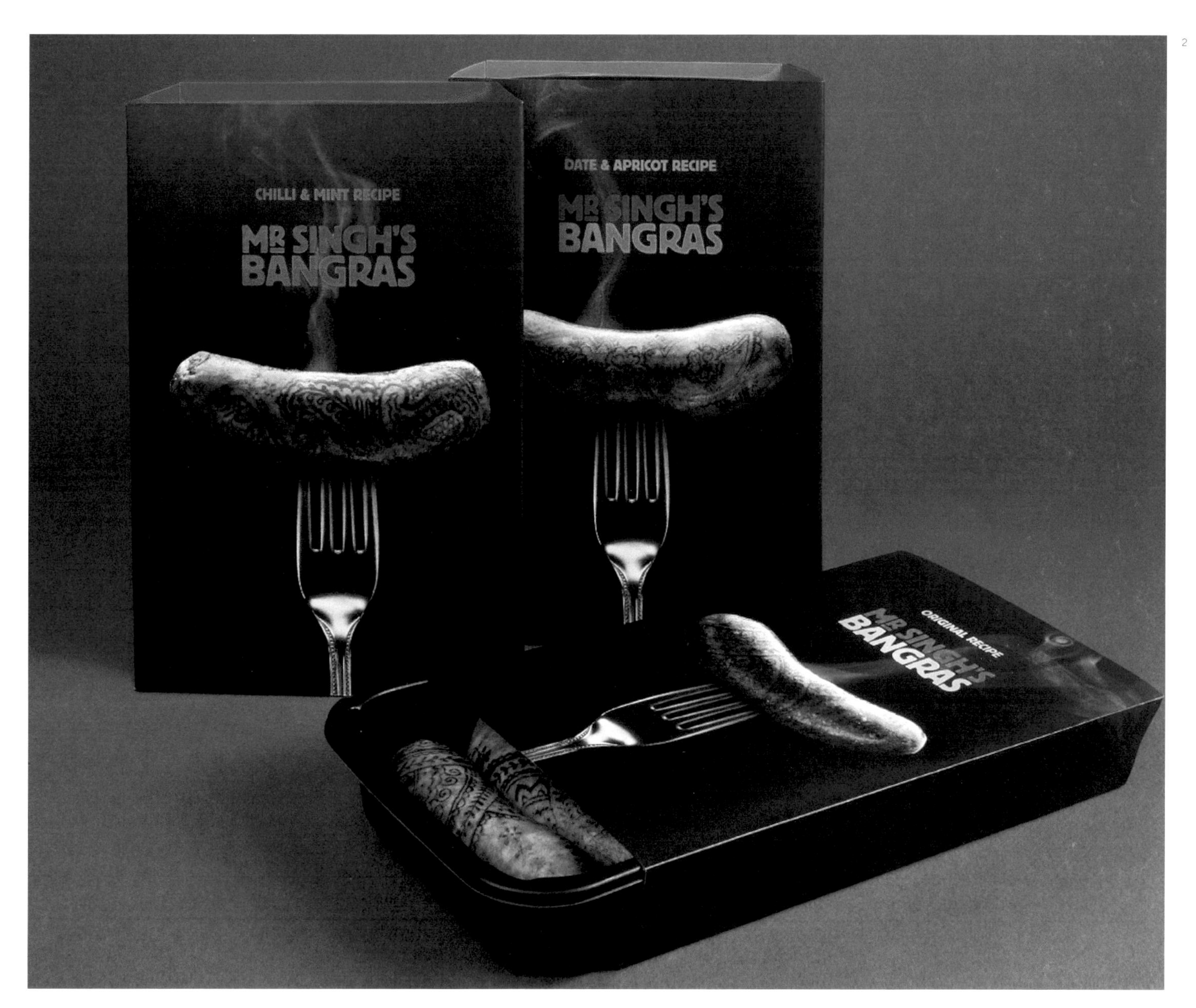
CHILLI & MINT RECIPE
MR SINGH'S BANGRAS
DATE & APRICOT RECIPE
MR SINGH'S BANGRAS
ORIGINAL RECIPE
MR SINGH'S BANGRAS

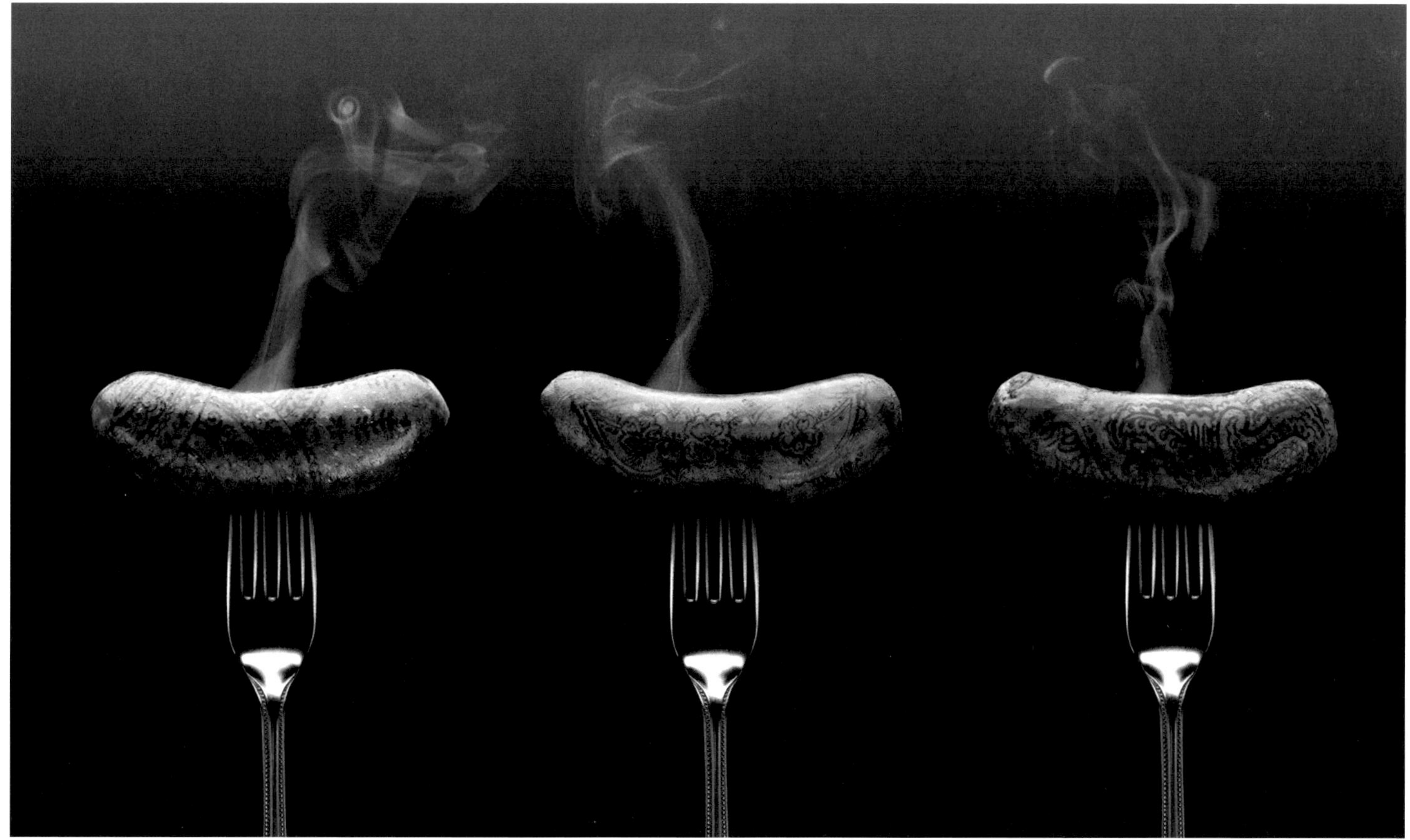

1

2

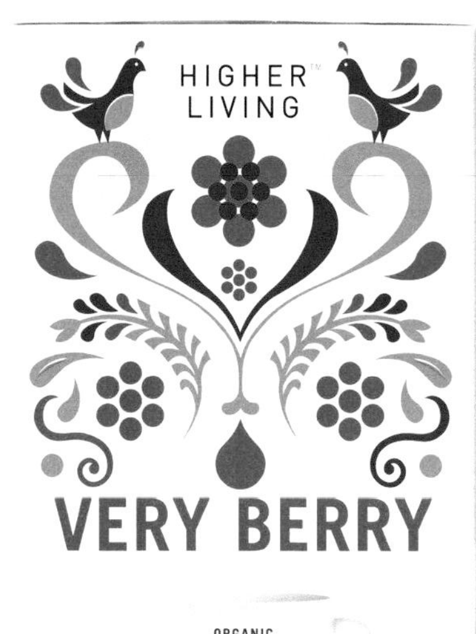
HIGHER LIVING
VERY BERRY
ORGANIC
HERB INFUSION
CAFFEINE FREE
15 BAGS

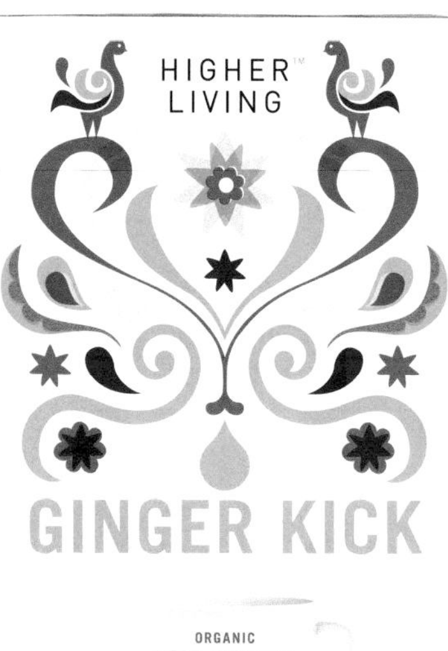
HIGHER LIVING
GINGER KICK
ORGANIC
HERB INFUSION
CAFFEINE FREE
15 BAGS

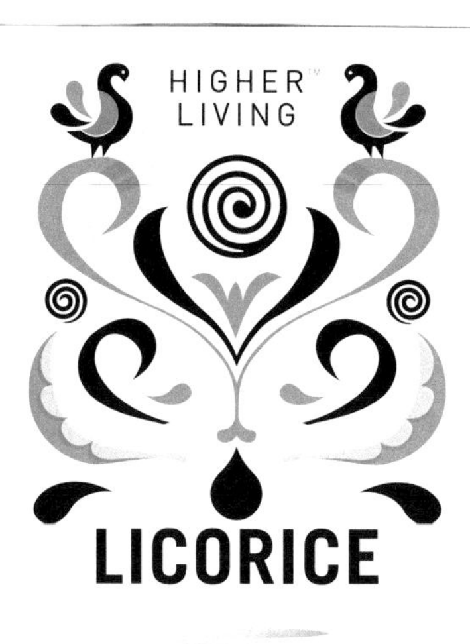
HIGHER LIVING
LICORICE
ORGANIC
HERB INFUSION
CAFFEINE FREE
15 BAGS

HIGHER LIVING
ORGANIC
HERB INFUSION
CAFFEINE FREE
15 BAGS

HIGHER LIVING
EVENING
ORGANIC
HERB INFUSION
CAFFEINE FREE
15 BAGS

HIGHER LIVING
VITALITY
ORGANIC
HERB
INFUSION
15 BAGS

HIGHER LIVING
CINNAMON
ORGANIC
HERB INFUSION
CAFFEINE FREE
15 BAGS

HIGHER LIVING
COOL MINT
ORGANIC
HERB INFUSION
CAFFEINE FREE
15 BAGS

MATTMO

1 Product: World Expo Shanghai, Food Packaging Dutch Pavilion, Client: World Expo Shanghai, Distribution: China, 2010
The food packaging developed for the Dutch pavilion of the World Expo 2010 in Shanghai offers a modern, illustrative twist to traditional Dutch iconography while maintaining a low carbon footprint.

B&B STUDIO

2 Product: Higher Living Teas, Client: Higher Living, Distribution: UK, 2011

STAN HEMA

3 Product: Ökohof Die Kuhhorster, Client: Die Kuhhorster (Mosaik WfB gGmbH), Distribution: Germany, Since 2009

3

COUPLE

1 Product: The Hand Burger, Client: The Soup Spoon, Distribution: Singapore, 2009

Inspired by the client's philosophy of serving bespoke burgers, the identity developed was likewise handcrafted: all packaging elements are crafted entirely from paper, and aim to highlight the artistry of making a good burger. The comprehensive product array runs the gamut from burger containers, labels and picks to cups and cupholders, which are harmonized by a customized alphabet made from scratch—like the burgers they represent.

WILLIAMS MURRAY HAMM

2 Client: Blue Skies, Distribution: Ghana, UK, 2010

Reflecting the fruit company's sustainable production cycle, and its motto "from field to fork," the packaging graphics use the pipeline as a motif throughout the product range. Various stations along the "fruit pipeline" are illustrated in a narrative of simple graphic elements, depicting the cycle that goes from the source, to transport, to product consumption, and finally back to the farms.

ILOVEDUST

3 Product: Bill's Milkshakes, Client: Bill's Dairy Farm, 2009

A fun, candy shop-style packaging for this range of milkshake assortments was developed to appeal to a young target market of 12 to 25-year-olds. Each flavor features its own character and specific color palette.

MAYDAY

Barry Gillibrand, Roger Akroyd

4 Product: Fruit and Herb Vinegars, Client: Womersley Fine Foods, Distribution: UK, 2010, Material: paper label applied to glass bottle, with black vinyl neck capsule

MINT DESIGN

5 Product: Holiday Packaging, Client: Adidas, Distribution: USA, 2007

LUIS TRÉPANIER

6 Product: Ataka, student-project, 2011

Packaging for this brand of Canadian-grown cranberries named for the Iroquois word for the fruit uses a traditional Native American pattern, while its color evokes both the fruit, and the Canadian flag. The package was designed to be used to its full potential: a strip is torn out to open the package, separating it into two serving bowls.

1

FREE MINI STRAWBERRY MILKSHAKE WITH EVERY PURCHASE OF THE HANDBURGER ORIGINAL
RAFFLES CITY SHOPPING CENTRE, B1-77/78

THE HAND BURGER
MENU
THE HANDBURGER ORIGINAL $13.8
TEA-SMOKED DUCK BURGER $15.8
CHICKEN CAESAR BURGER $12.8
TANDOOR CHICKEN BURGER $13.8
PULLED PORK BURGER $11.8
PARMESAN PRAWNCAKE BURGER $17.8
BEER BATTERED DORY BURGER $14.8
STUFFED PORTOBELLO BURGER $13.8
VEGETARIAN KAKIAGE BURGER $10.8
THE WORKS BURGER $16.8
EVERYTHING ELSE
DRINKS AND DESSERTS

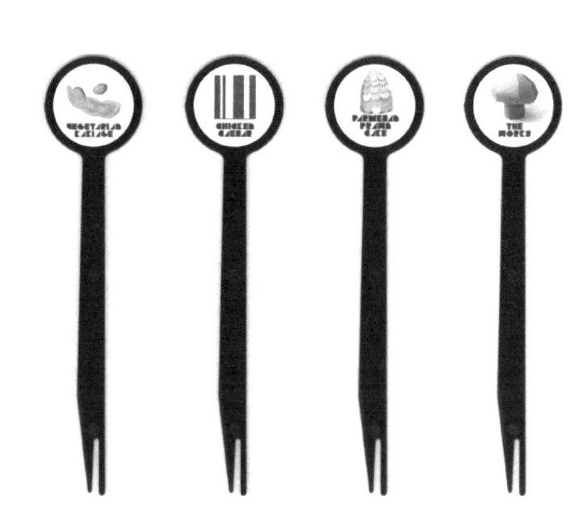

3

BILL'S
DAIRY FARM
VANILLA MILK

BILL'S
DAIRY FARM
CHOCOLATE MILK
NATURE'S BEST

BILL'S
DAIRY FARM
STRAWBERRY MILK

2

SUGARLOAF PINEAPPLE
TROPICAL FRUIT
HANDPICKED POMEGRANATE
COCONUT CHUNKS

BLUE SKIES
FRESHLY SQUEEZED
PINEAPPLE & GINGER JUICE
BLUE SKIES
FRESHLY SQUEEZED
TROPICAL JUICE
BLUE SKIES

4

WOMERSLEY
BLACKBERRY VINEGAR
WOMERSLEY
CHERRY VINEGAR
WOMERSLEY
WOMERSLEY
RASPBERRY VINEGAR
WOMERSLEY
BLACKCURRANT & ROSEMARY VINEGAR

5

4/5
4/5
4/5

6

ATAKA
BISCUITS SABLÉS AUX
CANNEBERGES ET AMANDES

AIAIAI

Peter Mix Willer/Kilo-Lars Holme Larsen

1 Product: Swirl and Pip, Client: AIAIAI, Distribution: worldwide, 2007–2009

ANDREI POPA

2 Product: Magni, self-initiated, 2009

1

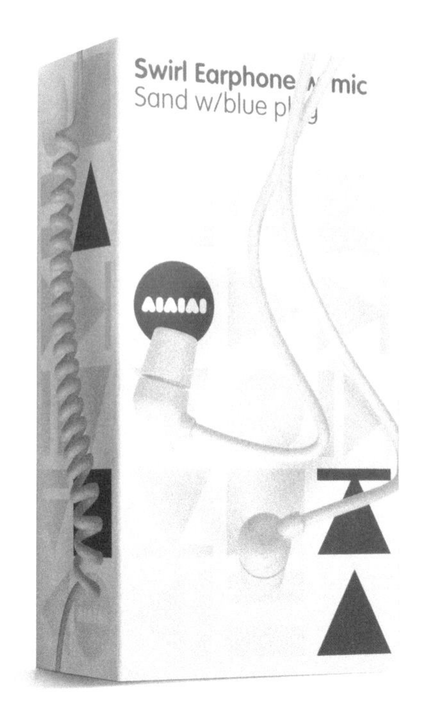

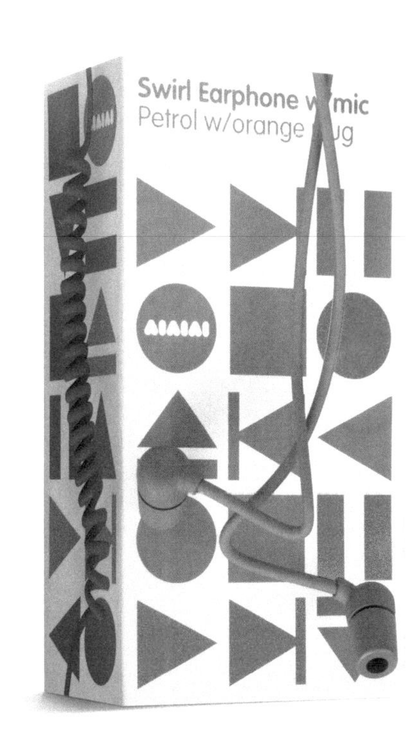

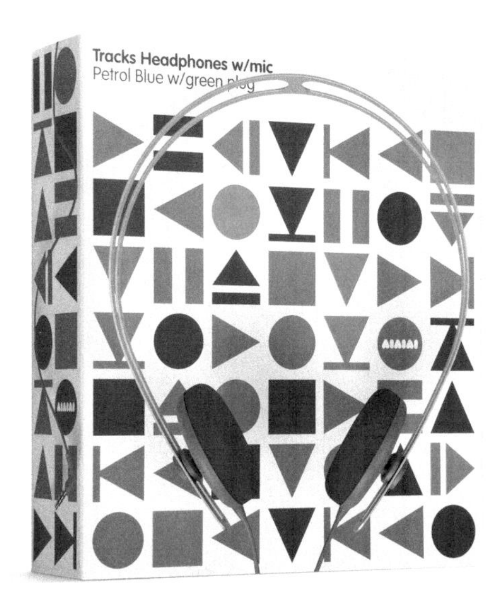

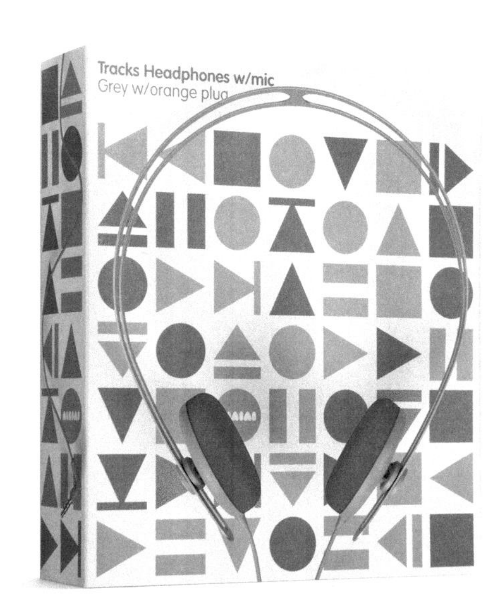

2

MAGNI
chocolate & honey
100g e

MAGNI
chocolate & milk
100g e

MAGNI
white chocolate & honey
100g e

AMORE
Product: Nature's Paper, Client: Tetra Pak,
Distribution: worldwide, 2009

Peak
Freshness
Pure
Simplicity
Tetra Fino Aseptic
In school, at home, or miles from anywhere, Tetra Fino Aseptic brings pure simplicity to packaging liquid dairy products for people to enjoy.
Senses
TPA 500 ML SQUARE
STREAMCAP
IN SHAPE
FOR MORE

AMORE

1 Product: Nature's Paper, Client: Tetra Pak, Distribution: worldwide, 2009

In order to showcase Tetra Pak's primary material, packaging print motifs were produced by photographing paper in an array of colors and forms: pleated, scrunched, flat, torn, piled, shredded, and even as confetti. Careful attention was taken to select images evoking both emotional and functional aspects of the products, and which would fit all of the company's packaging structures both within a "family" range, and for stand-alone products.

RIC BIXTER

2 Product: Elastic Bands, self-initiated, 2011

MEETA PANESAR

3 Product: Op Art Wine, self-initiated, 2009

This graphically driven packaging project brings together the art of winemaking with visual art, taking its design inspiration from the colors and patterns of Joseph Albers, and the op art movement.

1

2

ONE HUNDRED
Premium Quality
EXTRA TENSILE
EXTREMELY THIN
LIGHT
YET SURPRISINGLY TOUGH
ELASTIC BANDS

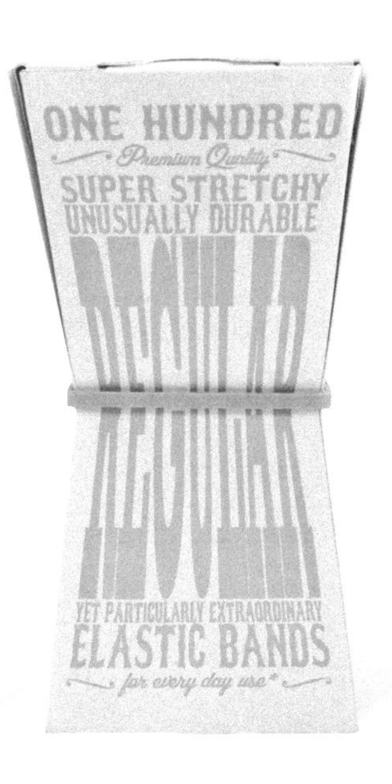
ONE HUNDRED
Premium Quality
SUPER STRETCHY
UNUSUALLY DURABLE
REGULAR
YET PARTICULARLY EXTRAORDINARY
ELASTIC BANDS
for every day use

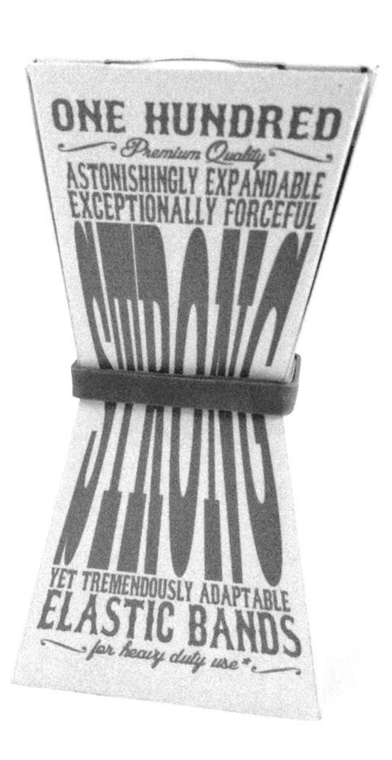
ONE HUNDRED
Premium Quality
ASTONISHINGLY EXPANDABLE
EXCEPTIONALLY FORCEFUL
STRONG
YET TREMENDOUSLY ADAPTABLE
ELASTIC BANDS
for heavy duty use

3

MERLOT
albers winery
MERLOT
albers winery
CABERNET
albers winery
CABERNET
SAUVIGNON
albers winery
SHIRAZ
albers winery
SHIRAZ

THAT WAS THEN...
THIS IS NOW
THAT WAS THEN...

YVES BÉHAR | FUSEPROJECT

Client: Jawbone, Distribution: Japan, China, Korea, Russia, Turkey, Saudi Arabia, Spain, Italy, Germany, Latin America, Brazil, France, Canada, Australia and online, 2010

Creating an emotional connection with the consumer was the goal of this packaging concept. While the exterior of the box recalls eighties music culture with a silkscreened boom box graphic, and the words "that was then..." printed on the lid, the inside reveals the Jawbox speaker tagged with "this is now," thus underlining the evolution of both culture and product innovation. The 100 percent recyclable packaging uses die-cut accessory cavities, and a snug middle tray instead of plastic ties or protective material, and furthers the company's sustainability goal.

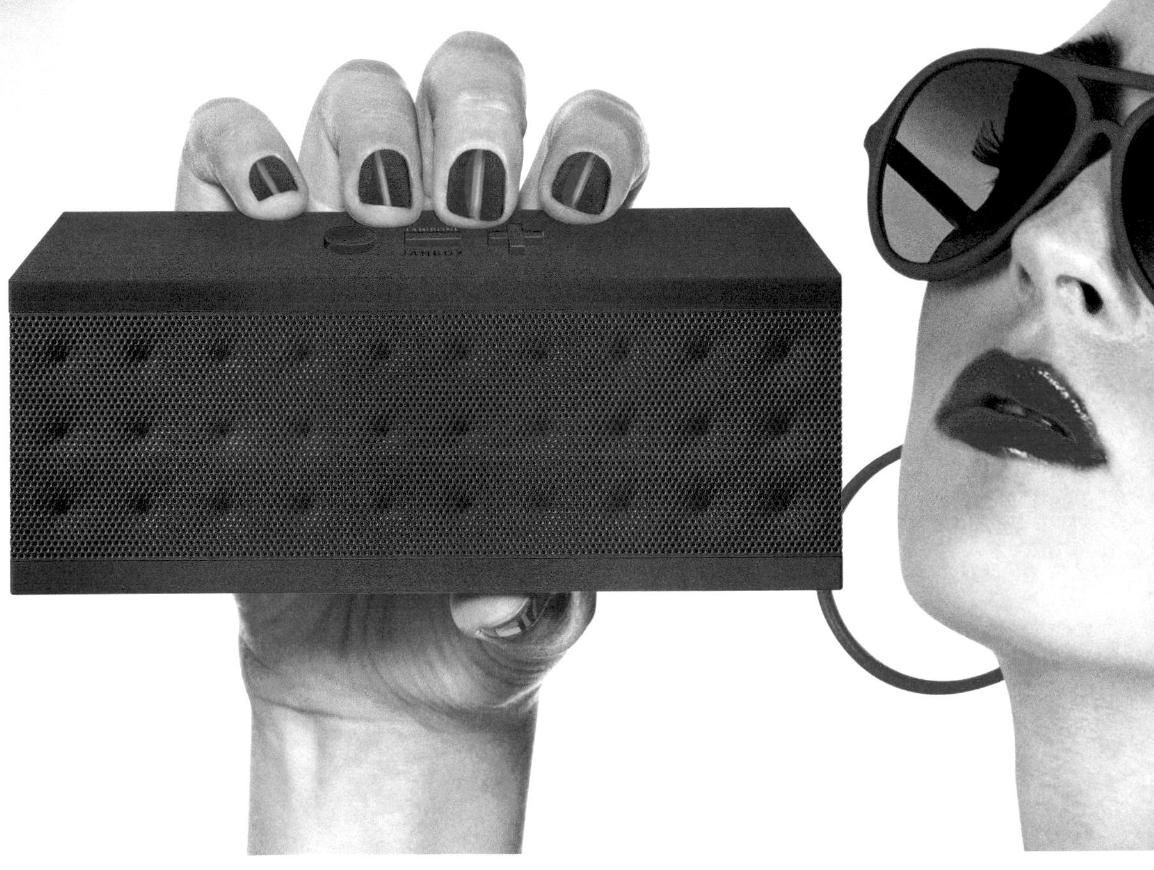

**NORRA NORR
AND ZOUND INDUSTRIES**

1 Product: Minor/Major, Client: Marshall Headphones, Distribution: worldwide, 2010, Material: cardboard

The packaging gives a nod to the brand's heritage while adding functionality of product display for the retail environment. The rough exterior of the box is reminiscent of the traditional makeshift packaging of amplifiers. When opened, the floor of the inner packaging elevates and tilts towards the viewer for an enticing show of the Marshall headphones.

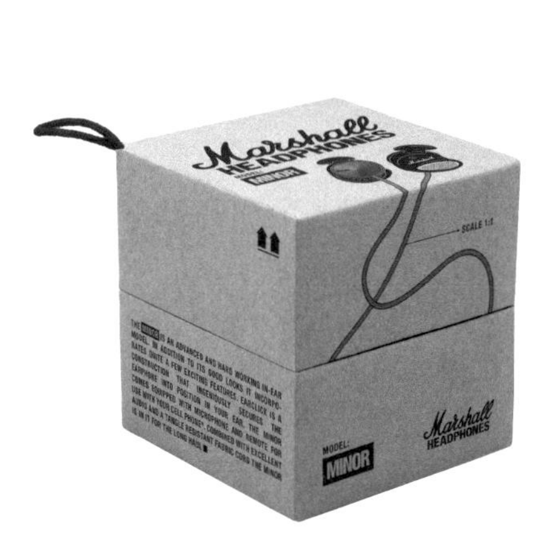

1

NORRA NORR AND ZOUND INDUSTRIES

2 Product: Urbanears, Client: Urbanears, Distribution: worldwide, 2010, Material: Cardboard

The packaging for this range of headphones aims at maximizing visual impact in the retail context, facilitating recyclability, and creating a memorable unpacking experience for the consumer. Boxes come in 14 different shades to follow the color concept of the products, while opening the package is like unfolding an origami creation.

noté
Écouteurs stéréo pour sports
Stereo earphones for sports.
noté
Écouteurs stéréo insonorisés.
Sound isolationg stereo earphones

CORINNE PANT

Product: Note Earbuds, student-project, 2010, Material: single piece of carboard with no gluc

Noté earbuds are the result of a student project to find an alternative cost and environmentally considerate packaging for a notoriously over-packaged product. Forsaking the use of glue or plastic, the earbuds are wrapped around a single piece of carton to simulate musical notes, thus becoming the main communication elements of the packaging.

ALVVINO (ALESSANDRO MAFFIOLETTI)

1 Product: Crumpled City Maps, Client: Palomar S.r.l., Distribution: worldwide, 2010

SARAH DERY

2 Product: Dr. Mullets, self-initiated, 2010

1

2

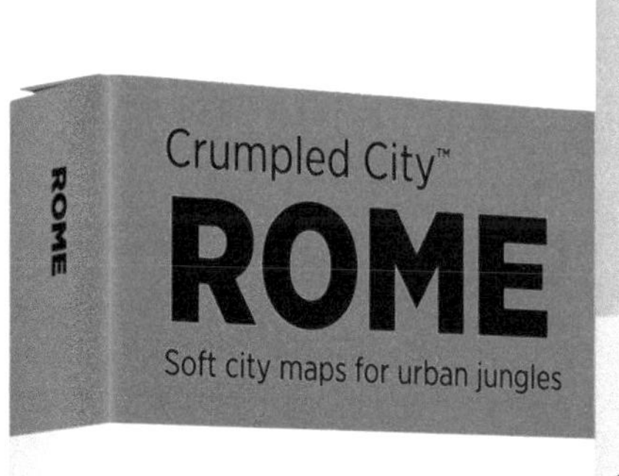

ROME
Crumpled City™
ROME
Soft city maps for urban jungles

PARIS
Crumpled City™
PARIS
Soft city maps for urban jungles
Crumpled City™

PLAIN HEAD
HATE KNOTS
BAD CURLS
BORING HAIR

1

KOCHZEIT 8-10 MIN
THE DELI GARAGE
MEHRZWECKNUDELN
DÜBELNAGEL
SECHSKANTSCHRAUBE
SECHSKANTMUTTER
8mm
19mm
21mm
15mm
SERVIERVORSCHLAG
250 g Genuss
THE DELI GARAGE
WWW.DELI-GARAGE.COM

2

THE DELI GARAGE
STRACCIATELLA
BACK
STEINE
300g | 6 BACKSTEINFÖRMCHEN
Die lockeren Stracciatella-Backsteine aus Dinkelmehl gelingen mit den beigefügten Förmchen garantiert auch Menschen mit zwei linken Händen. So lecker, dass jeder gerne beim Reinschaufeln hilft.
REINES DINKELMEHL

THE DELI GARAGE
SCHOKO
BACK
STEINE
400g | 6 BACKSTEINFÖRMCHEN
Mit extra viel Schokolade schmecken diese feinen Backsteine großen Handwerkern genauso wie kleinen Nachwuchsbastlern. Einfach die integrierten Förmchen befüllen, ausbacken und zusammen wegspachteln.
100% GLUTENFREI

3

THE DELI GARAGE
SCH
OKO
LEIM

THE DELI GARAGE
SCH
OKO
LEIM
MIT KEKS

STRACCIATELLA
BACK
STEINE

THE DELI GARAGE
SCHOKO
BACK
STEINE
400g | 6 BACKSTEINFÖRMCHEN
100% GLUTENFREI

4

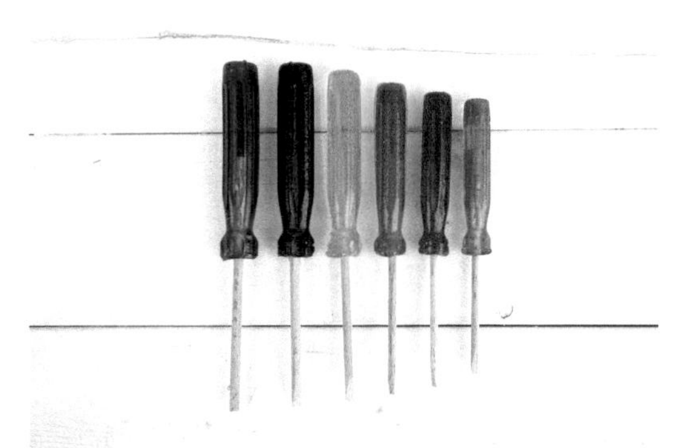

KOREFE

Client: The Deli Garage, Distribution: Europe, 2009–2011

The diverse range of products designed by Korefe for The Deli Garage is held together by the unifying notion of the garage, and the company's handcrafted foods. This is expressed in the use of a utilitarian typography throughout the range, along with diverse packaging evoking common repair shop tools, while the whimsical illustrations underline the playful creativity of the brand and its consumers.

1 Product: Multi Noodles

Pasta holds together many a meal—which is one of the reasons why Multi Noodles were crafted in the shapes of screws, dowels, nuts, and bolts. The toolbox packaging features clearly delineated compartments that ensure both the freshness and the organization of these essential culinary components.

2 Product: Brickstones

3 Product: Chocolate Glue

The white glue bottle is essential inventory in every workshop, which is why the Deli Garage decided to use it as packaging for its Chocolate Glue. Letters in the product's name come to life as a playful monster anticipating the added flavour to your favourite snack.

4 Product: Lolly Tools / The Deli Garage, 2011

The normal lollipop form was inverted for the playful Lollitool. Edible screwdriver handles come in six different flavours, while their (non-edible) white tips function as the lollipop stick. The limited edition treats are packaged in a classic leather tool bag, bringing together exclusivity with handicraft.

5 Product: Food Finish

Food Finish is sold in a spray can to evoke the standard dispensers of automotive lacquer, while the fanciful, hand-drawn illustrations on the cans bring together art and food.

5

KOREFE
Client: The Deli Garage, Distribution: Europe, 2009–2011

LACK
ROT
THE DELI GARAGE
BLAU

CHAPTER 2

MINIMAL & ELEGANT

In the early 1960s, the milkman still delivered door to door in my small hometown. The glass bottles were simply deposited on the front porch where my mother had left tokens corresponding to the number of quarts she desired. These beautiful glass bottles were striking in their immaculate whiteness, and in their total absence of graphics or branding. A few years later, this magnificent deposit system was dismantled and the bottled milk was replaced by powdered milk, which came packaged in paper sealed envelopes featuring a vaguely identifiable brand. We were told that the pure bottled product had been technologically simplified. It was supposedly just as good for us and, yes, much cheaper.

If the story of the great brands is marked by originality and distinction, the dominant trend in packaging today is characterized by a visual overkill often resulting in a standardization of graphic codes. If we can look to displays at any corner or grocery store, we will see how codes, colors, and visualization are similar from one brand to another. Here, the chameleon effect is particularly strong and brand distinction is almost lost. Similarly, nothing looks more like a brand of instant soup than another brand of instant soup? Indeed, take a closer look and across all the brands you will more or less find the same logo, in the same top left position, with the same photograph of a bowl, and the spoon ready for you to sip.

In addition to visual homogenization there is also a great deal of legal information such as nutrition information tables, an ingredient list as well as the various health claims—organic, sugarless, contains omega 3 and so forth. The package surface is an area increasingly optimized, maximized, and overloaded with visual content. The brand itself is ever more in competition with its own communications—lost in a chaotic and unreadable graphic environment. There is therefore a paradox between the desire to establish strong brands and the very real, and limited surface space of the package. In a bilingual market, as we have here in Canada, this space is further divided in two and the surface available to establish compelling communication is reduced to almost nothing. This information overload can sometimes be read as a sign of weakness, doubt, or apprehension with respect to the brand. If the brand is strong and has its own voice, why is there a need to prop it up with noisy discourse? If everyone speaks at the same time, the message is not heard. For this reason, a subtle but clear message is sometimes much more effective than the cacophony of the status quo. As simple a strategy as this may seem, it is one we rarely see on our shelves.

The minimal style is a response to the visual overkill phenomenon, which pervades across our visual culture. As with the functionalist theories that gave rise to it, the minimal style in packaging advocates an approach based on plain functionality and clear product identification. It is an approach that relies on the intelligence of the user and which enhances the strict front and center branding of the product. The minimal style values the subtle aesthetics of simplicity, materiality, and the product itself.

But if this approach has the merit of certain objectivity, it can also be monotonous and contravene the basic principles of distinction. It may suffer from a lack of personality. If the competitors decide to mimic the minimal style, the entire shelf can become a bland and featureless landscape. Devoid of all graphics and typographical excess, minimalism in packaging may— by its severity— be perceived as cold, elitist, and without soul.

The minimal style is perhaps a more intellectualized style of design that invariably circles back to Mies van der Rohe's now famous if not clichéd exertion that "less is more." I tried to explain this modernist notion to my young son the other day, but without success. For Felix, less is still just less. It's a reaction that cannot be taken lightly given that our fascination for packaging is an emotional experience that takes root in childhood. Indeed, there is a six-year-old child hidden within every consumer. Today, the minimal style may evoke different emotions than it has at other times in its history, but it remains a style used often for a targeted audience. It's coldness and its bareness tends to evoke a sense of elegance and refinement often associated with luxury products or customers perceived to be more educated and more sensitive to design.

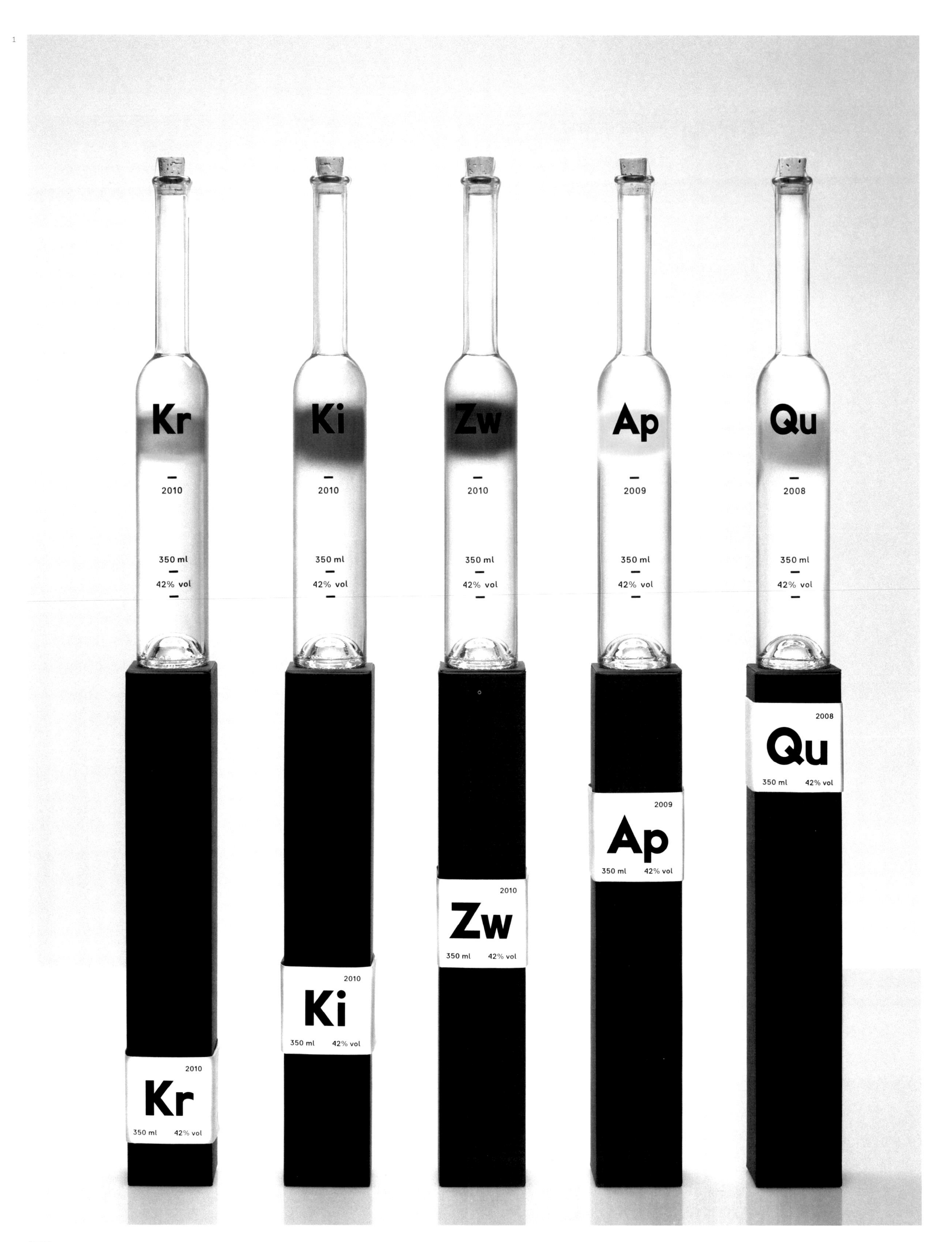
Kr
2010
350 ml
42% vol
Ki
2010
350 ml
42% vol
Zw
2010
350 ml
42% vol
Ap
2009
350 ml
42% vol
Qu
2008
350 ml
42% vol

THOMAS LEHNER

1 Product: Rezept-Destillate, Swiss Fruit Spirits, Client: Rezept-Destillate, Distribution: Germany Switzerland, 2010-2011, Material: Packaging: black cardboard; package band: lightface press paper. Bottle: industrial glassware, natural cork, adhesive film, colored spray

A simple and unique solution was found for this small range of fruit spirits that would both accommodate the limited budget as well as evoke an impression of high class, and exclusiveness. Minimalistic labelling and abbreviated naming on the clear glass bottles, and simple cartons recall the aesthetic of 19th century apothecaries and convey the producer's expertise.

ACME INDUSTRIES

2 Product: Batteries, Client: Volta, Distribution: Romania, 2010

AARON RICCHIO

1 Product: Wonder Daily, Client: Wonder Bread, Distribution: USA, 2010, Material: silkscreened white glass Boston round bottles with matching colored metal caps

NEUMEISTER

2 Product: Arla Milk Bottles, Client: Arla, 1998

Adding a modern twist to the traditional milk bottle, this packaging aims to position the product as a healthy alternative to other popular beverages.

1

2

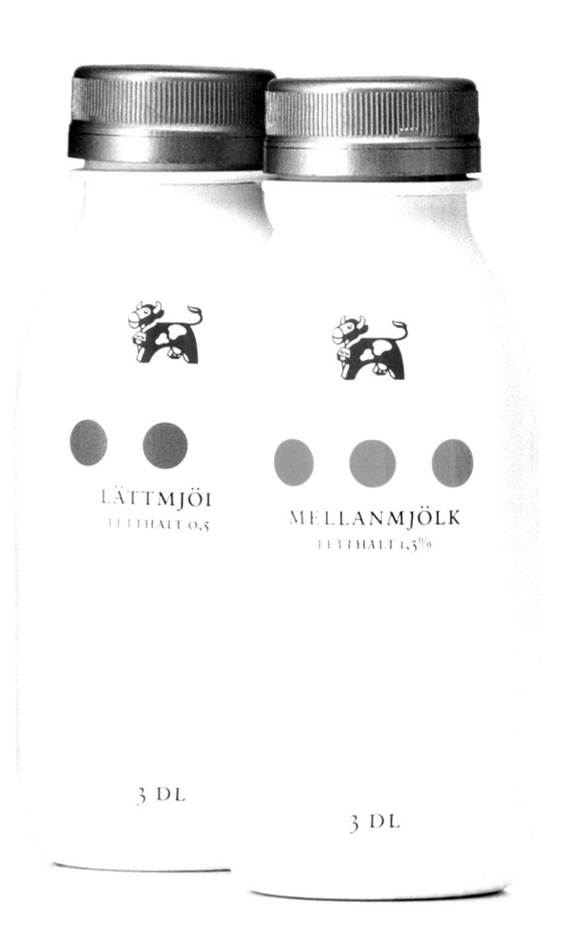

BASEDESIGN

3 Product: Pantone Fan Guide, Client: Pantone, Distribution: worldwide, 2010
The identity developed for the new Pantone product line started with the name "plus" to emphasize the new improvements to the previous system. The logo and icon were framed in a visual context based on the iconic Pantone Chip. The design concept applies to the packaging for the entire product line as well as digital, and print communications.

BOB HELSINKI

4 Product: KC Professional—Prism, Client: Miraculos, Distribution: Finland, 2010
The packaging concept for this range of hair color pigments uses the metaphor and form of the prism as its starting point to create a diverse range of colorful, and recognizable packaging solutions.

3

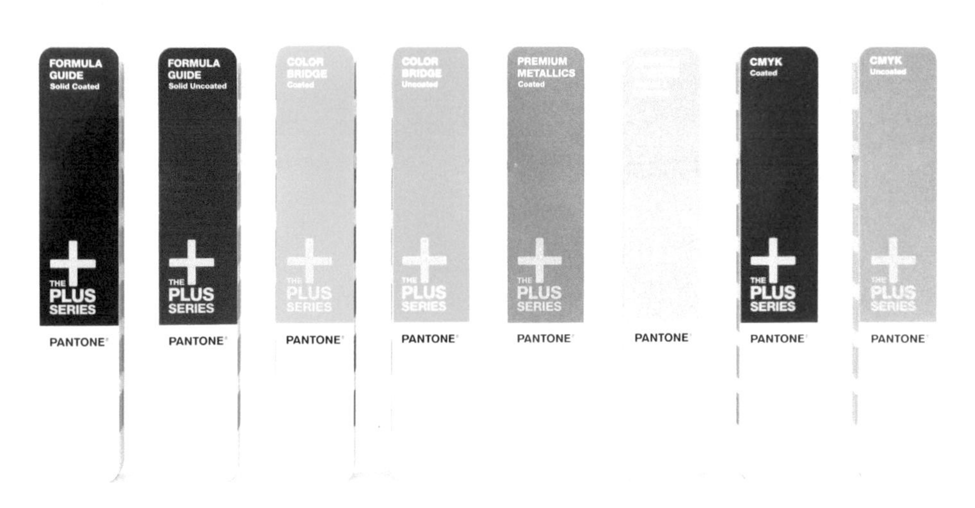

4

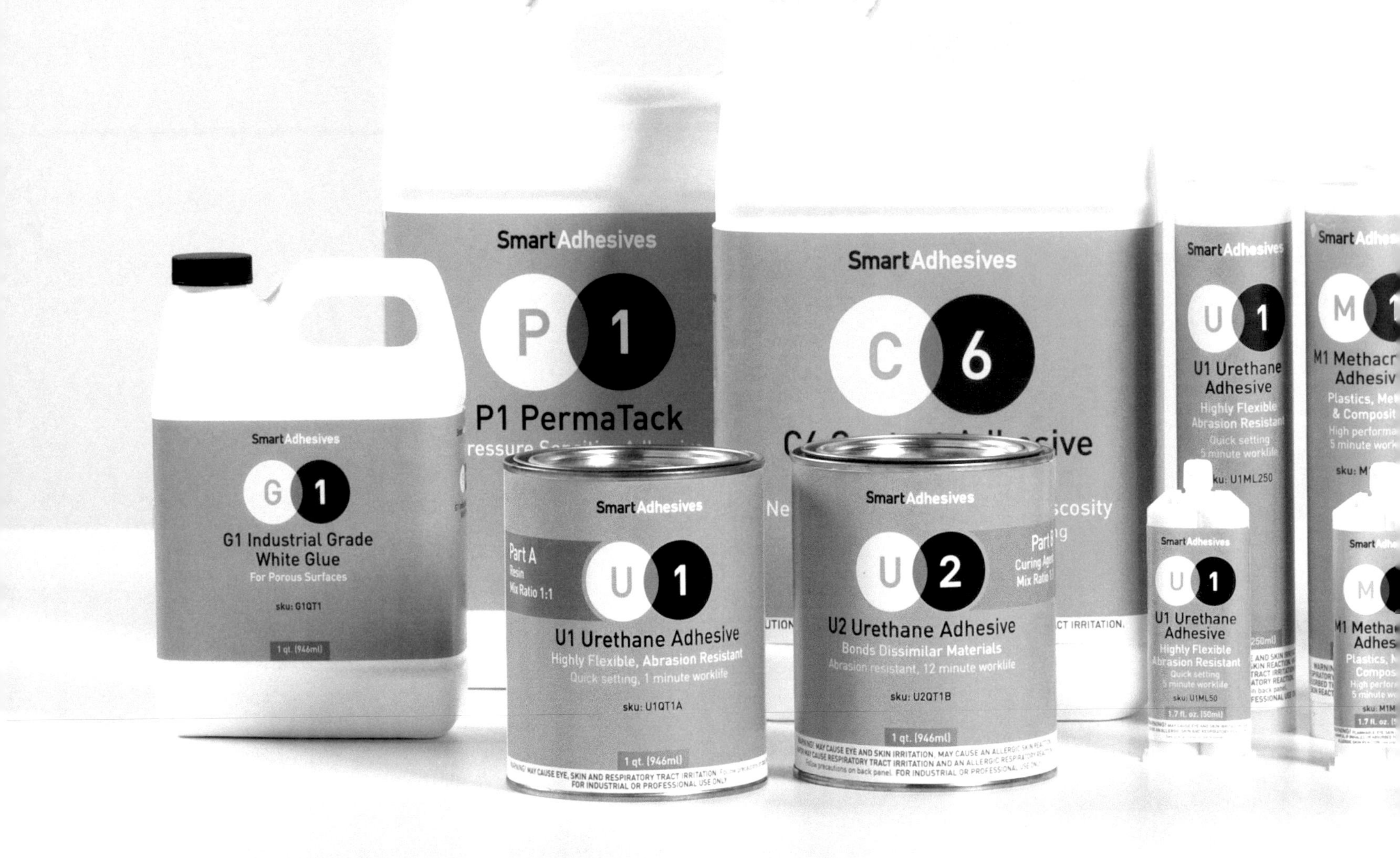

1

2

STANDARD STUDIO

1 Product: various products, Client: Smart Adhesives, Distribution: worldwide, 2009, Material: matte paper on several packages

STEPHANIE KUGA

2 Product: Chroma Paint, Client: The FreshAire Choice, Distribution: Concept, 2009, Material: e-flute, LDPE
The sustainable, easy to use modular redesign of this environmentally safe paint is inspired by photosynthesis, and suggests an eco-friendly image and reliability. The identity concept allows for the collection to grow as technology advances.

THE METRIC SYSTEM

3 Client: Steen-Hansen Maling, 2008

Smart Adhesives
C 3
C3 Contact Adhesive
Nitrile Formula
Heat & chemical resistant
sku: C3GL1
1 gal. (3.78L)
Smart Adhesives
C 4
sku: C4GL1
1 gal. (3.78L)
E 1
E1 Epoxy
Rigid, Quick Setting
1.7 fl. oz. (50ml)
i 4
Impact Resistant
sku: i4OZ16
16 fl. oz. (473ml)
i1 Instant Adhesive
T 4
T4 Threadlocker
T 3
Threadlocker
High Torque, Permanent
General purpose
sku: T3ML250
8.4 fl. oz. (250ml)
T 1
T1 Threadlocker

3

Mester
maling
Oljebasert husmaling
Steen-Hansen
Mester
dekk
ELITE V
Palett
Mester
Mester
dekk
Mester
maling

The Hub
BIKE RESCUE PROJECT
PUNCTURE REPAIR KIT
INNER TUBE
SPANNER SET

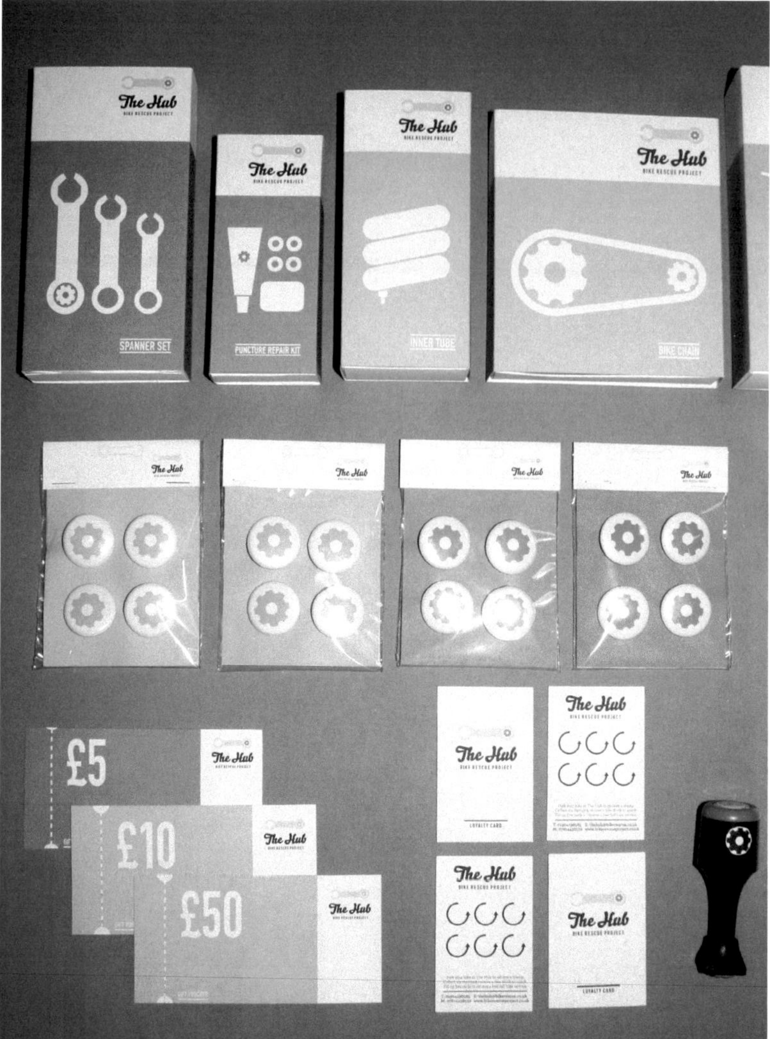
The Hub
BIKE RESCUE PROJECT
SPANNER SET
PUNCTURE REPAIR KIT
INNER TUBE
BIKE CHAIN
£5
£10
£50
LOYALTY CARD

1

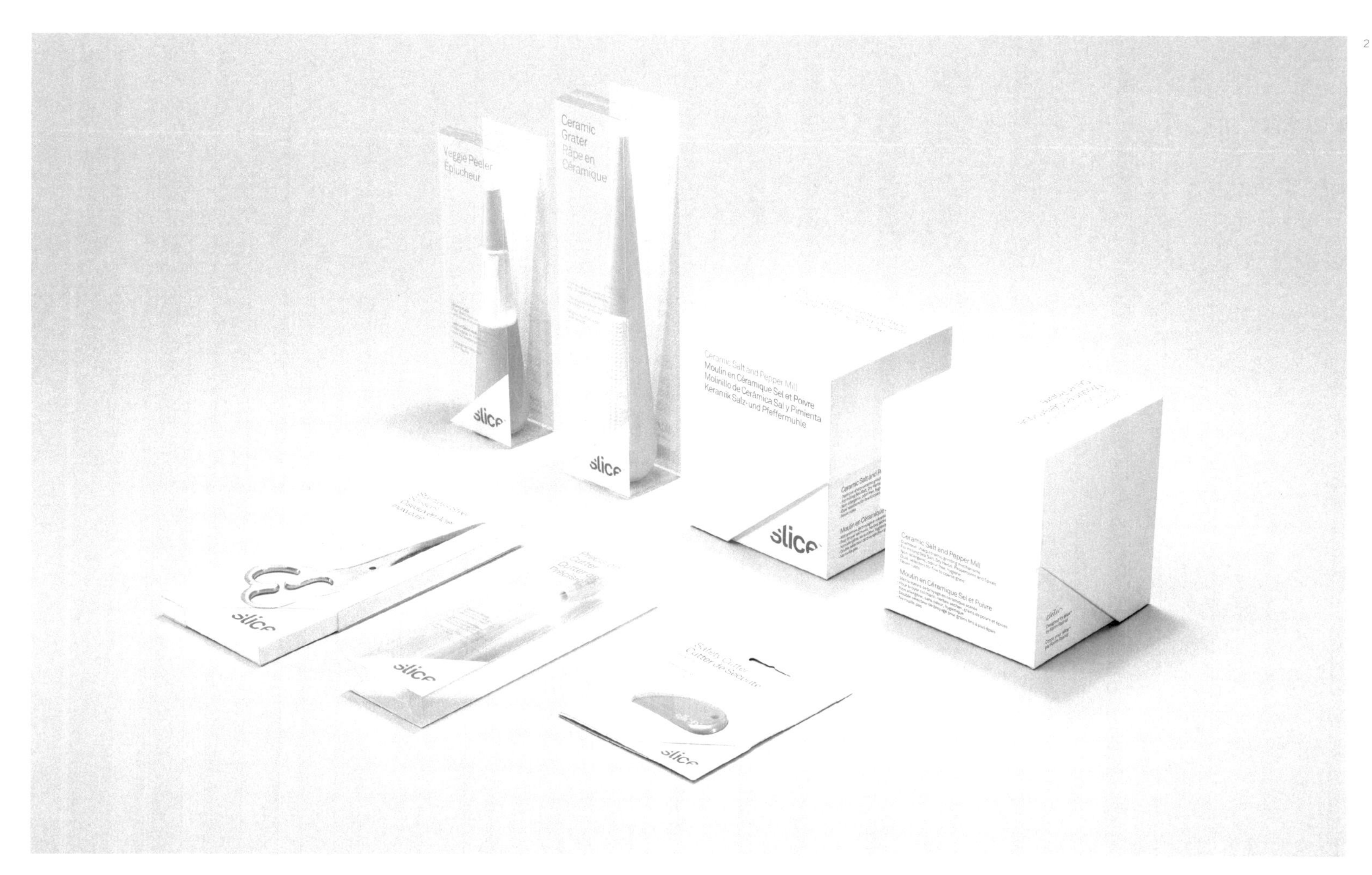
Veggie Peeler
Eplucheur
Ceramic
Grater
Râpe en
Céramique
slice
Ceramic Salt and Pepper Mill
Moulin en Céramique Sel et Poivre
Molinillo de Cerámica Sal y Pimienta
Keramik Salz- und Pfeffermühle

2

3

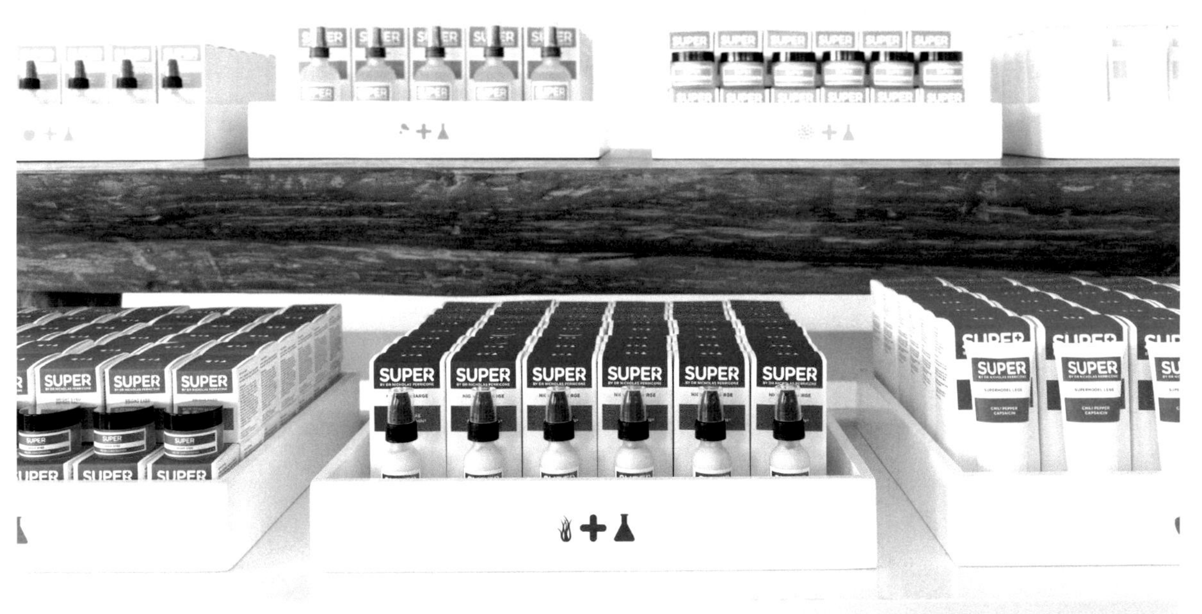

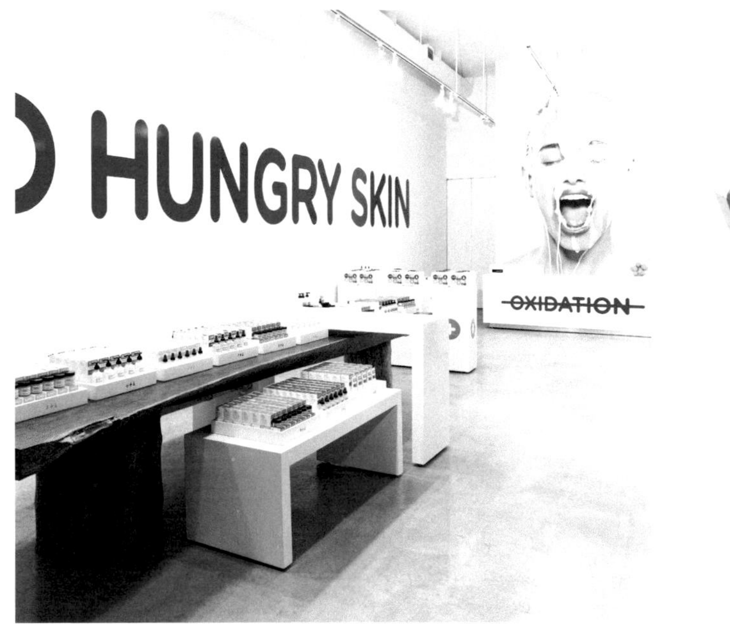

DAVE RAXWORTHY

1 Client: The Hub - Bike Rescue Project, 2011

MANUAL

2 Client: Slice, Distribution: USA, Europe, 2010, Material: paper over rigid board construction

An iconic brand logo was created by literally slicing into the word "slice," in order to instantly communicate the nature of the company, and its range of innovative cutting tools. This angle becomes an integral part of the packaging system, often being used architecturally to reveal typographic information, and to add a physical branding element to the structures.

CONCRETE DESIGN COMMUNICATIONS

3 Product: Super by Dr. Nicholas Perricone, USA, 2010

The identity developed for this range of dermatological products is based on the notion of nutritional "superfoods," a term coined by the company's founder. The packaging, which uses 100 percent recycled material and simple one-color printing, combines a sense of fun, and whimsy with information on the science of the products to target a broad, and young customer base.

ESPLUGA+ASSOCIATES

1 Product: Mesoestetic Christmas Greetings 2009, Material: Client: Mesoestetic, Distribution: Spain, 2009

Simple packaging makes bold use of typography and color to quickly convey the intention of this Christmas chocolate edition.

NOAH BUTCHER

2 Product: Eighthirty Coffee Roasters, Client: Eighthirty, Distribution: New Zealand, 2009, Material: compostable coffee cups

The combination of a clear and simple design matched with cheeky messages printed on the coffee cups and packaging aim to stand out in an already cluttered market.

PENTAGRAM DESIGN LIMITED

Angus Hyland

3 Product: Cass Art Bags, Client: Cass Art, Distribution: UK, 2009, Material: nonwoven fabric

An artist supply store promotional tote bag takes a typographic approach to link the store's inventory with great works of art.

1

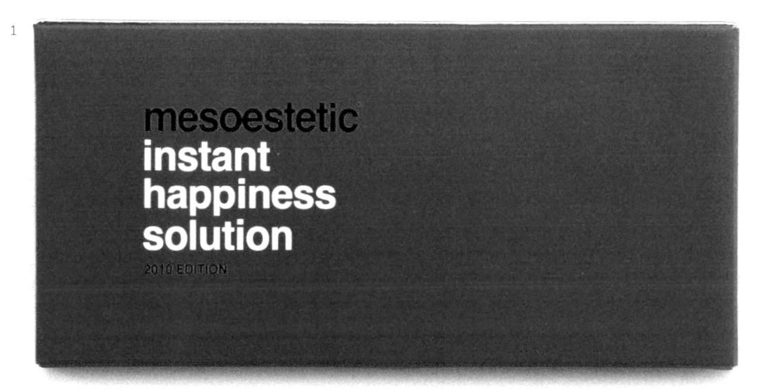

2

CASS ART LONDON
SCARLET
LAKE
WWW.CASSART.CO.UK

CASS ART LONDON
RUFINO TAMAYO (1899–1991)
WOMAN WITH RED MASK
NICOLAS DE STAËL (1914–1955)
RED BOTTLES
EMIL NOLDE (1867–1956)
RED POPPIES
JOHN MARTIN (1789–1854)
THE GREAT DAY OF HIS WRATH
JAMES ROSENQUIST (1933—)
STUDY FOR MARILYN
PALMA VECCHIO (C1480–1528)
THE HOLY FAMILY WITH MARY MAGDALENE AND THE INFANT SAINT JOHN
HENRI MATISSE (1869–1954)
HARMONY IN RED
ALEXEI VON JAWLENSKY (1864–1941)
SCHOKKO
DALER ROWNEY

CASS ART LONDON
PHTHALO
TURQUOISE
WWW.CASSART.CO.UK

CASS ART LONDON
TAMARA DE LEMPICKA (1898–1980)
THE TWO FRIENDS
GIOGIO DE CHIRICO (1888–1978)
SONG OF LOVE
DAVID HOCKNEY (1937—)
CALIFORNIA BANK
CHRIS OFILI (1968—)
PRINCE AMONGST THIEVES
TERRY WINTERS (1949—)
LUMEN
ARTHUR DOVE (1880–1946)
LAKE AFTERNOON
ERNST LUDWIG KIRCHNER (1880–1938)
GROUP OF ARTISTS
MAXFIELD PARRISH (1870–1966)
WINTER NIGHT
WILLEM DE KOONING (1904–1997)
THE VISIT
DALER ROWNEY

TRAN HUYN
1 Product: Gi, Client: Omega, Distribution: USA, 2010

LAINEY LEE
2 Client: Chroma, 2010
Inspired by the form of a lipstick tube, when the cosmetics bag is not held, its handle is hidden inside the bag. The handle rises when the bag is picked up, mimicking the shape, and movement of lipstick lifting out of its tube.

SKIN DESIGNSTUDIO AS
3 Client: Laid AS/Laid , Distribution: Norway, USA, Canada, Australia, UK, Germany, France, Netherlands, Italy, Lithuania, Estonia, Denmark, Hungary, Czech Republic, Greece, China, Russia, 2010

CAITLYN GIBBONS
Student redesign of Kmart, Distribution: USA, 2009
4 Product: K Salt, Epson HW matte
5 Product: K Lightbulb
The redesign project aims to better position Kmart with an in-house economy range of food, home, health, and beauty as well as an organic food line. The packaging throughout the range is succinct, utilitarian, and recyclable.

1

2

D.1
Laid
Silicone Dildo
D.1
Laid

P.2
P.2
P.2 Silicone Cock Ring
P.3
Laid
P.2
Laid
Laid
P.3 Silicone Cock Ring
D.1

P.2
Silicone Cock Ring
Laid
P.2 Silicone Cock Ring
P.2

3

4

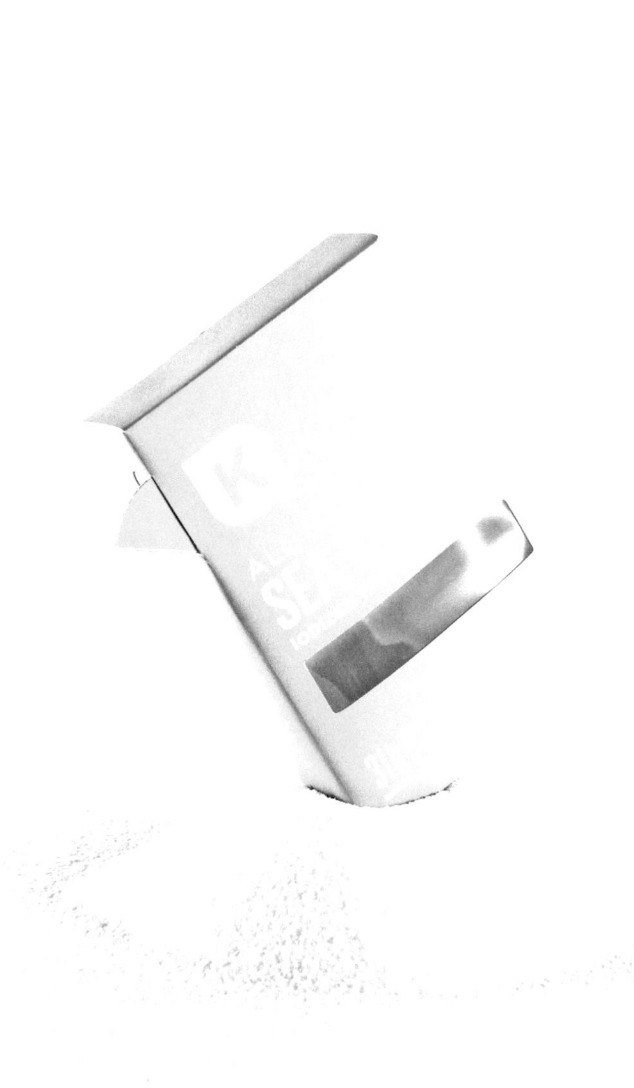
K

5

K
LIGHT BULB
COMPACT FLUORESCENT
23

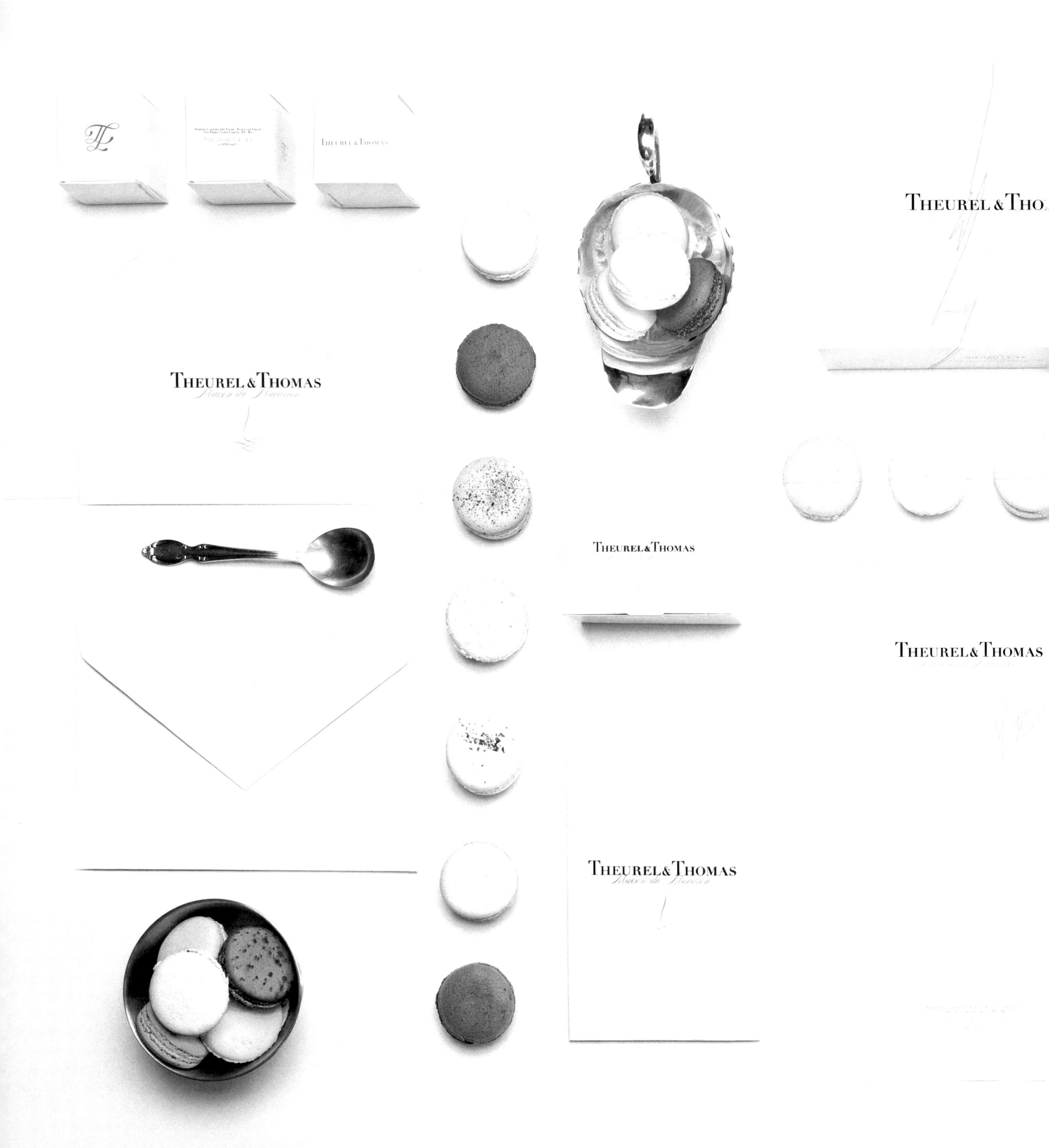
THEUREL & THOMAS
THEUREL & THOMAS
THEUREL & THOMAS
THEUREL & THOMAS
THEUREL & THOMAS
THEUREL & THOMAS

ANAGRAMA
Client: Theurel & Thomas, Distribution: México, 2009
A sophisticated brand identity was created for this French-style macaron patisserie in Mexico. The Didot typeface is used in signage, labelling, and communications, while two lines of cyan and magenta against a white background serve as a discreet, and avant-garde reference to the French flag.

A-B-D

1 Product: #1–#14, Client: Delights Of Sweden, Distribution: Scandinavia, 2010
The clear brand concept developed for the newcomer seasoning company easily accommodated the quick launch of a wide range of products while establishing a unique visual identity that stands out from its competitors on the shelf.

SHINE LTD.
with George Goldsack

2 Product: Jed's Coffee, Client: Bell Tea New Zealand, Distribution: New Zealand, 2010

ALT GROUP

3 Product: The Vigneron Centenary Wine, Client: Oakfield Press, Distribution: New Zealand, 2009
The packaging for the limited edition wine was designed to have a tactile quality, like the tools of the blind vigneron that it commemorates. The box is made of rough sawn timber and branded with his signature. In reference to the blackboard on which the winemaker would write his poetry, each of the 100 bottles was dipped in blackboard paint, sealed with wax, and then hand numbered with chalk.

94WINES

4 Client: 94Wines, Distribution: Netherlands, Belgium, Germany, 2009, Material: sleeved glass bottles

1

2

3

98/100
99/100
100/100

4

1
white

2
white
smooth

3
red
fruity

4
red
gentle

5
rosé
fresh

22
white
spicy

23
red
full

24
red
oaky

41
prosecco
sparkling

42
rosé brut
classy

51
white
luscious

52
red
rich

1

SAL LAGOS
ROSAS
Sal
Tres
Pimientas

SAL LAGOS
ROSAS
Sal
con
Hibisco

SAL LAGOS
ROSAS
Sal
Hierbas
Provenzales

2

EDUARDO DEL FRAILE

1 Product: Sal Con Circulos, Client: Sal Lagos Rosas, Distribution: Spain

2 Product: Queso de Cabra, Client: BEEE, Distribution: Spain, 2007

For this line of cheese and yoghurts made from goat's milk, bold typography is used to draw attention to the brand while the clear lines of the packaging, and the playful iconic logo make the products both recognizable, and practical for presentation on the shelf. The musicality of the name—reflecting the bleating of the goats—was inspired by a visit to the biodynamically run farm, where the melodies of Mozart resounded over the loudspeakers as the goats were being milked.

delishop
Limoncello
Limoncello

ENRIC AGUILERA ASOCIADOS

Client: Delishop, 2007

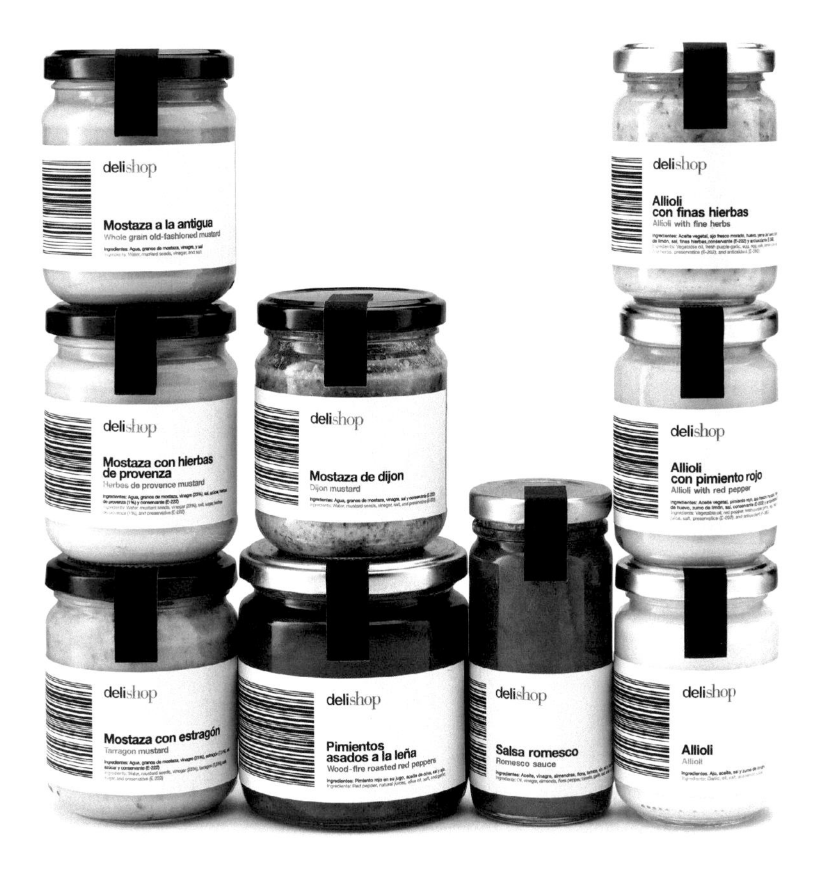

1

WODKA
WANESSA
PRODUCT OF AUSTRIA
DISTILLED FROM PURE SPELT
WITH ALPINE SPRING WATER
ALC 40%VOL
700ml

2

ORGANIC
Honey

ORGANIC
Honey

3

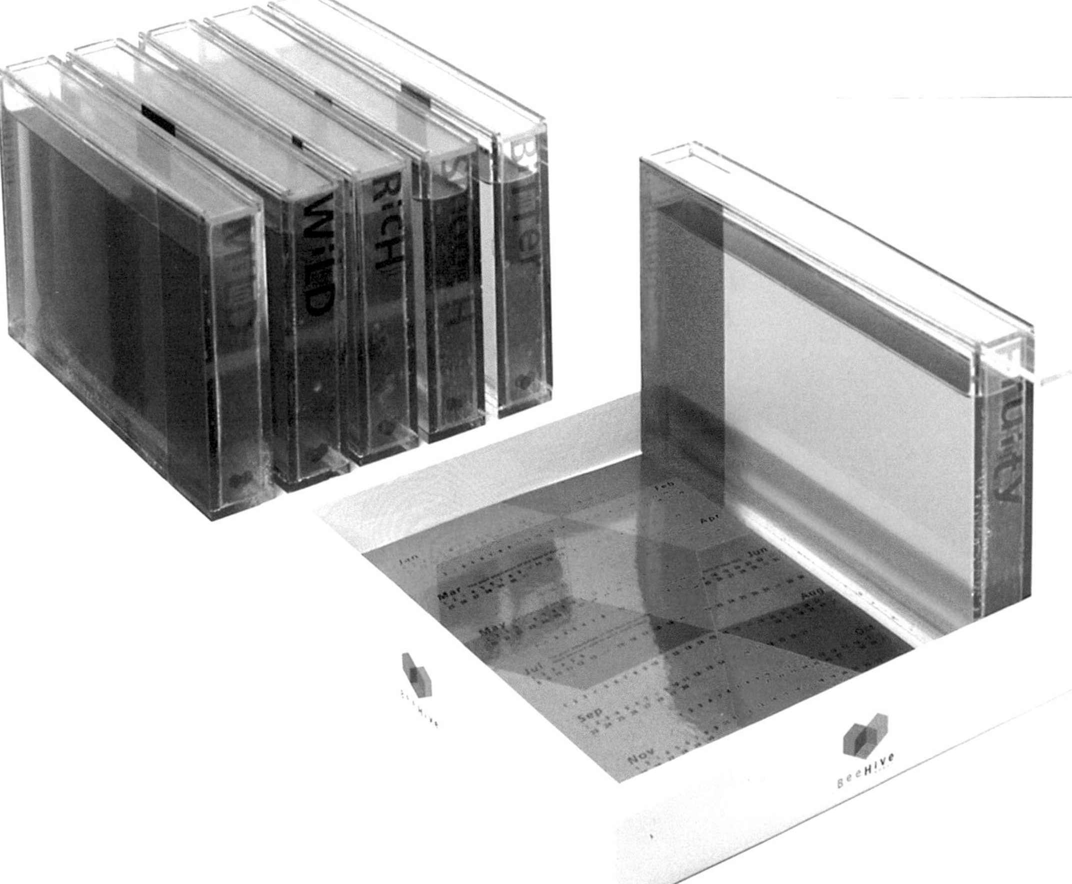

WODKA WANESSA
1 Product: Wodka Wanessa, Wanessa Fruit Shot, 2011

MARCEL BUERKLE - CIRCUM PUNKT DESIGN
2 Product: Organic Honey, self-initiated, 2009

LITTLE FURY
3 Product: BeeHive Honey, Client: BeeHive Honey Store, Distribution: concept project, 2006, Material: Acrylic plastic, wood, and rub down

BALLARD BEE COMPANY/ TKTJ DESIGN
4 Product: Bh Honey, Client: Ballard Bee Company, Distribution: USA, 2010

4

1

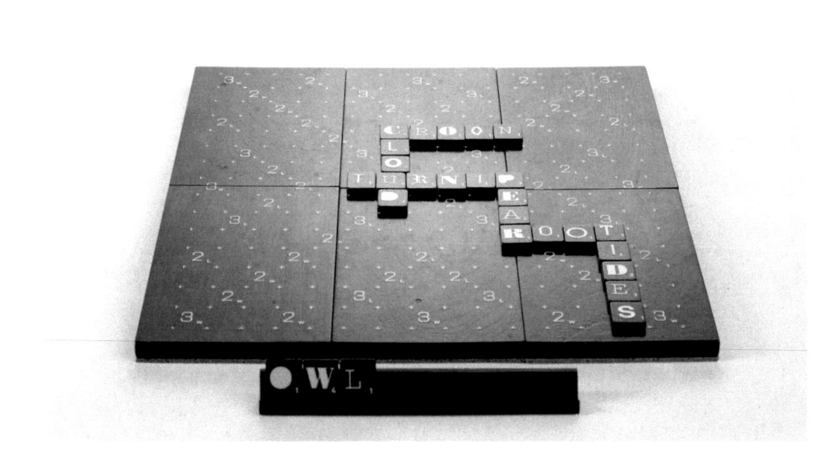

2

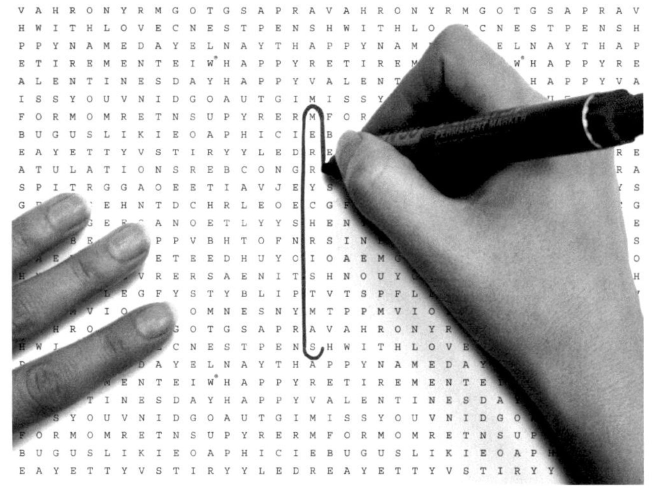

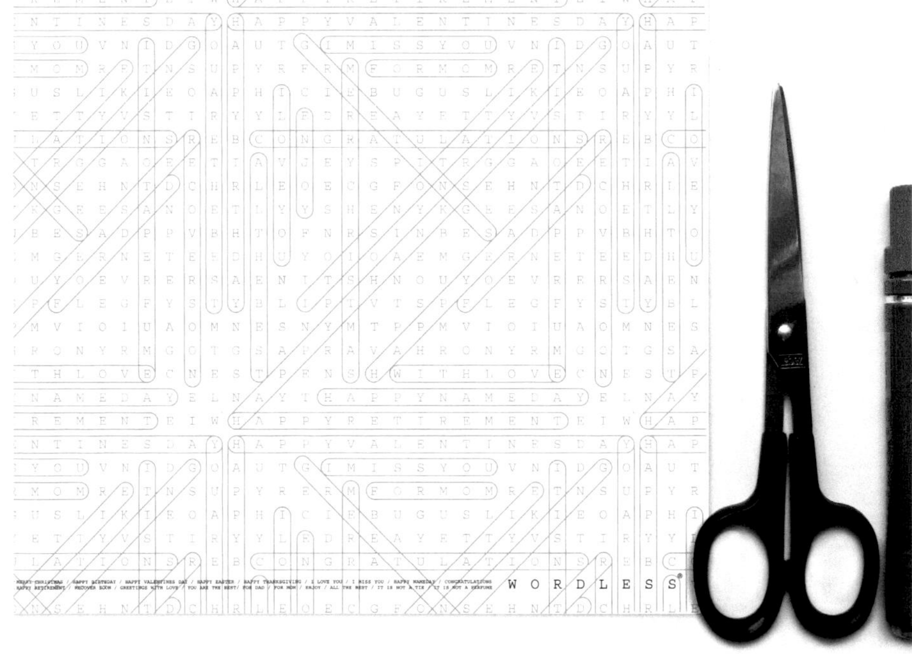

ANDREW CLIFFORD CAPENER

1 Product: A-1 Scrabble Designer Edition, designed during senior year at BYU

This redesign project interpreted the board game Scrabble as a commercially viable, designer's dream project. Letters in various font collections enable the player to create words in his or her favorite fonts, while additional font packs are available for future purchase. The game is stylishly packaged so that the individual board panels can be easily assembled with the help of magnets.

FABIO MILITO DESIGN

Fabio Milito, Francesca Guidotti

2 Product: Universal wrap, Client: Wordless design, Distribution: worldwide, 2009, Material: uncoated writable paper

Taking its inspiration from word search games, this writable wrapping paper conceals 20 different celebratory occasions that can be circled as needed, replacing a classic gift card. The word search solutions are printed on the inner side of the wrapping paper for quick reference.

baked beans
gives you a little bit more...
54%
MORE*
*for your money on average against the leading brand

2

RUTH PEARSON

1 Product: Value Range, student-project
The minimally designed labelling informs the consumer as to how much more you get for your money compared to leading brands.

BRIAN PASCHKE

2 Product: Cube Cloths, 2010, Material: uncoated stock with matte varnish
Compressed cleaning cloths are packaged into boxes the size of an ordinary dish sponge to emphasize their miniaturization. The white cube graphic that wraps around the box conveys the actual size of the product, while the color of the box indicates the quantity of its contents. Minimal type, and use of infographics make this product internationally friendly.

THE STANDARD

3 Product: The Standard Bath Amenities, USA, 2009
Keeping in line with this hotel's trade-mark no-frills, yet high quality ambi-ence, the Standard hotel's organic bath amenities come packaged in a practical zip travel case printed with thematically relevant pictograms. The regular branded commercial packag-ing of the Kiss My Face personal care line has been exchanged for a more "standardized" look and feel.

3

B
O
L
S
BOLS DECENT VODKA
SINCE CLOGS 750ML

MASH

Product: Bols Vodka, Client: Bols 1575, 2009

Heritage and the classic elegance were at the center of this pitch concept, developed for the relaunch of a vodka produced by one of the oldest distillery brands in the world. The proposal included a variety of bottles inspired by the look and feel of vintage French perfume packaging. Unfortunatly, it was ultimately not selected for use.

ARULIDEN

Product: PUMA PT3 Ultramagnetic Collection, Client: PUMA, Distribution: limited distribution in USA, 2007, Material: ABS plastic, EVA foam

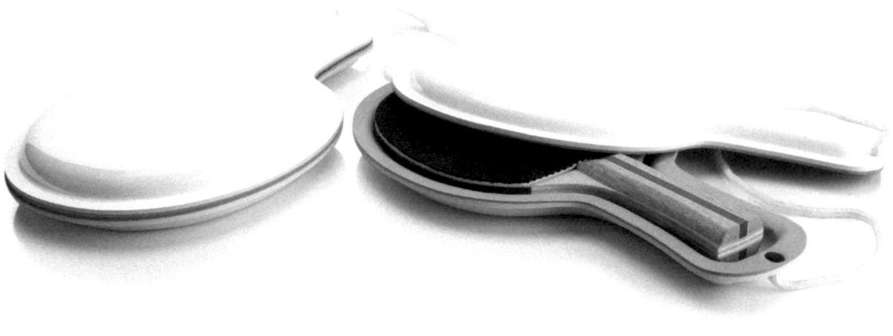

TAKU SATOH DESIGN OFFICE
Product: P.G.C.D., Client: P.G.C.D.Logi,
Distribution: Japan, 2000

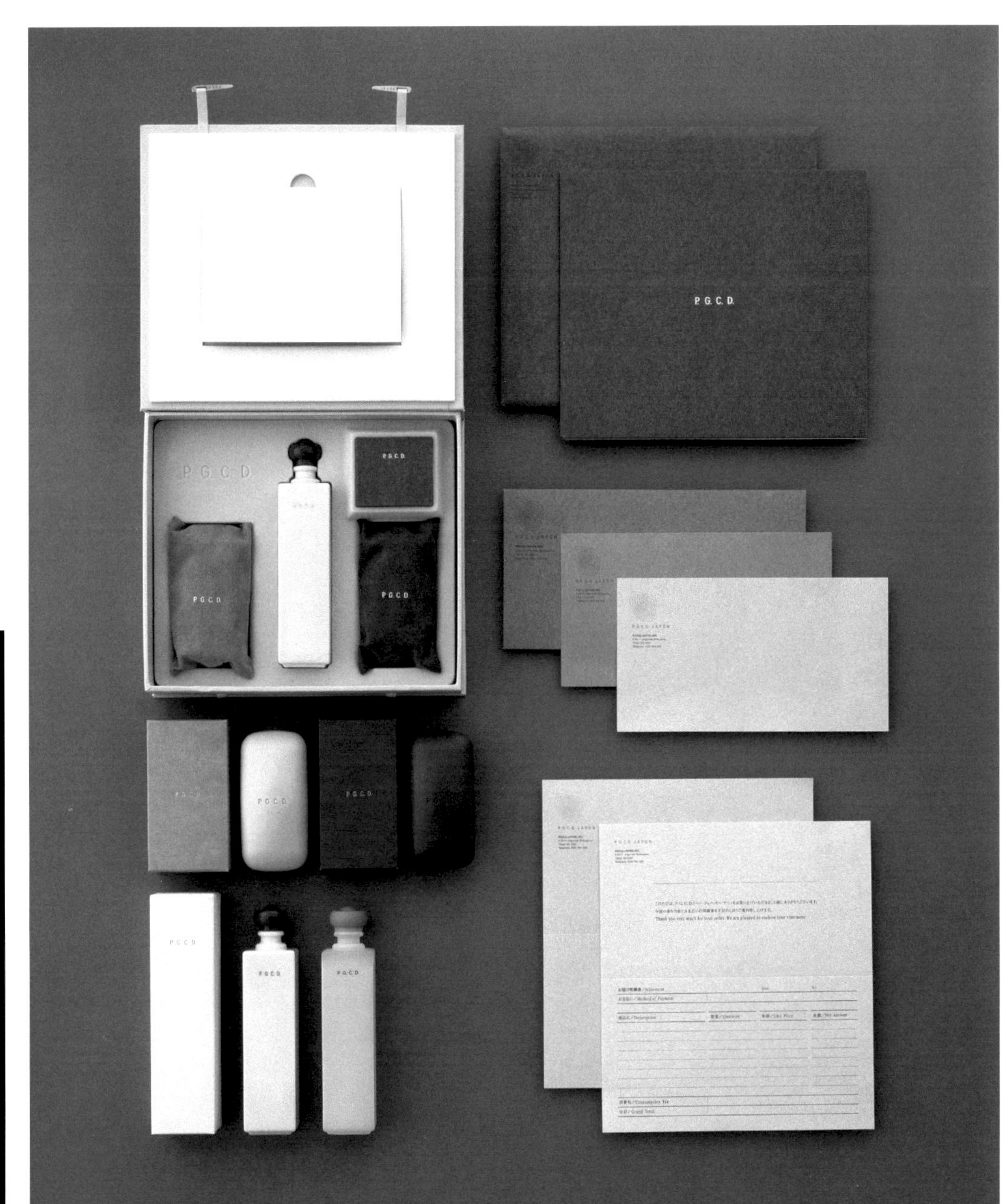

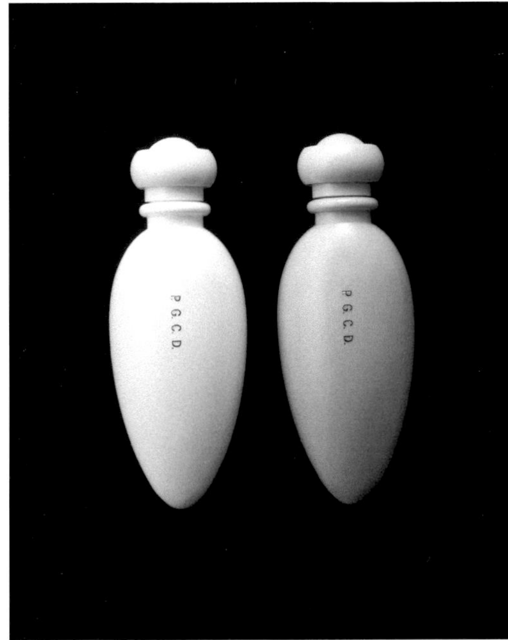

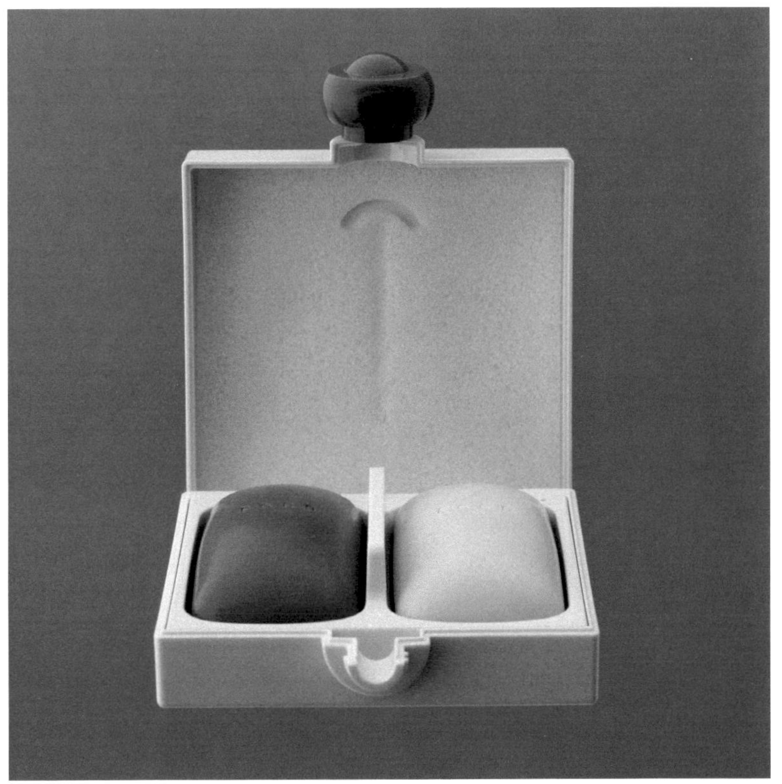

ABSTRACTIO
ВАСИЛИЙ КАНДИНСКИЙ
КОМПОЗИЦИЯ VIII

PETER ZHARNOV
Product: Abstractio, self-initiated, 2010,
Material: paperboard, aluminum foil

1

DEMIAN CONRAD

1 Product: Gabbani Packaging, Client: Gabbani, Distribution: Switzerland, 2010

NARANI KANNAN

2 Product: Neocube, Client: Neocube Ltd, Distribution: Online, 2009, Material: cardboard
Minimal packaging offers meditative inspiration for the myriad of shapes and patterns that can be composed with this high-energy magnet device.

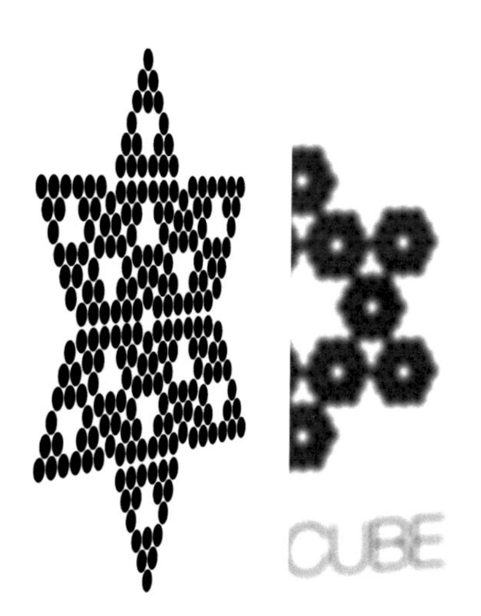

2

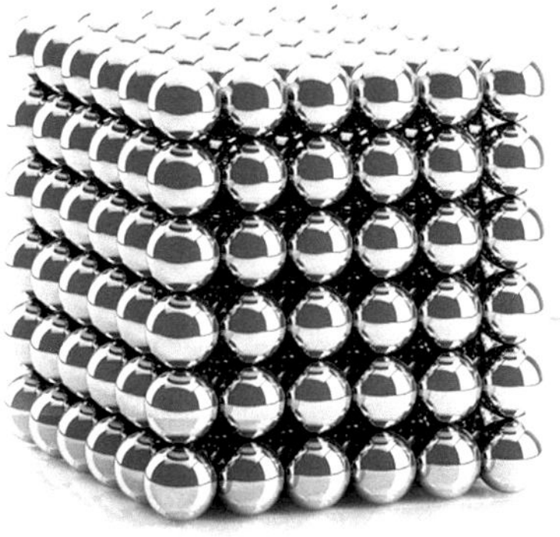

NEOCUBE

ALOOF

3 Product: White Scents Range, Client: The White Company, Distribution: UK, 2010, Material: uncoated paper over board, and folding boxboard

4 Product: Noir Range, Client: The White Company, Distribution: UK, 2010, Material: coated paper over board

3

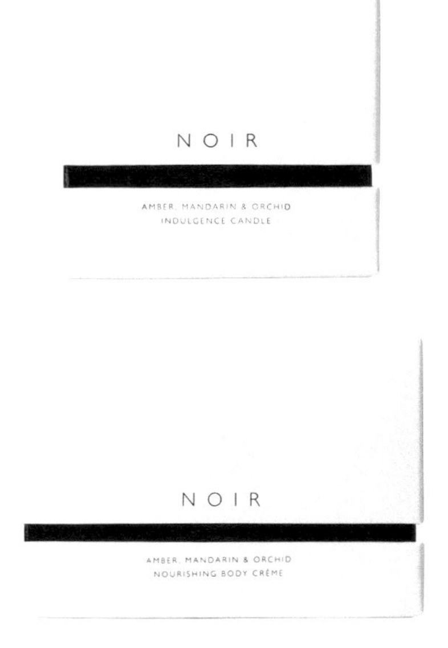

4

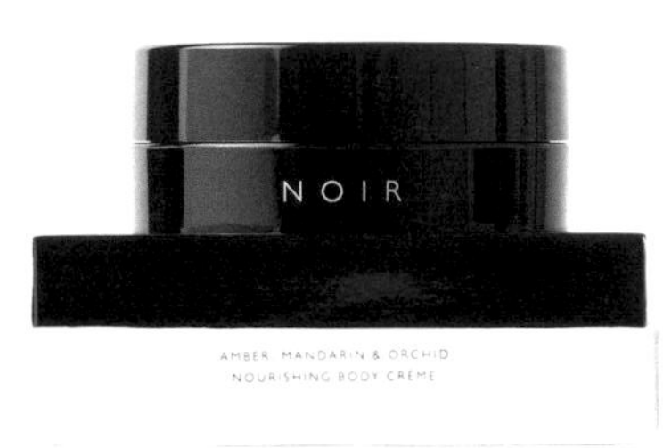

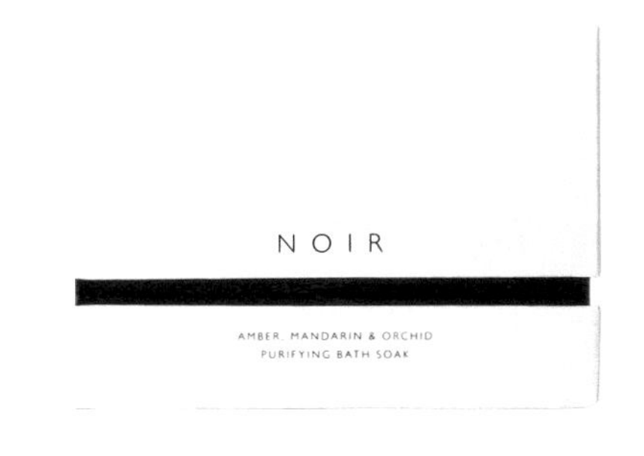

1

2

ALT GROUP

1 Product: A Lean Year, Client: Alt Group, Distribution: New Zealand, Australia, UK, 2009/2010, Material: black card (Notturno) box, foam die-cut inner

To celebrate the end of a year considered one of the worst in economic history characterized by a global financial slump, the bottles for this limited edition wine were themselves slumped, and then filled and corked by hand.

EDUARDO DEL FRAILE

2 Product: Pass-o, Client: Menhir Salento, Distribution: Europe, USA, 2010

3 Product: Extenso, Client: Agapito Rico

The packaging for this limited edition, celebratory wine was designed as an object of value unto itself, bearing an intrinsic value as a collectible reminder of the occasion upon which the wine was served.

3

NUTHENLUND
KNUTHENLUND

CHAPTER 3

Every fall, my grandmother would pay us a visit and share her autumn harvest with us. This ritual marked the end of summer and announced the approach of the harsh Canadian winter. I remember how the vision of those multicolored and wax-sealed glass jars filled with peppers, rhubarb, apple jelly, and pumpkin soup, were reassuring to me. Just as in earlier pioneering times—a time that I had not experienced—these colorful masonry jars seemed to hold the promise that we would be carried through the cold winter in comfort. Each jar was carefully labeled with the beautiful handwriting that my grandmother had been taught by nuns. The attention paid to each container inspired confidence and respect. The fruit of grandmother's labor promised us a unique and exclusive experience. At the end of winter, we would return the empty jars so she could refill them the next fall. This ritual ended when she died and from then on, mostly for practical reasons I suppose, we turned to the supermarket for our canned foods.

With the emergence of self-service sales in supermarkets, consumers have become less informed and less aware of the origins and processes behind the products they consume. Their modern and accelerated lifestyles no longer allows them to concern themselves with such questions, yet many will feel a certain anxiety, even cynicism, with respect to their consumption. This phenomenon might partially explain the nostalgic fascination for a bygone era. A picturesque past where proximity seemed to ensure a certain measure of quality or at least a sense of commitment between the producer and the customer.

With the Down to Earth approach, the quality of the product will often match the claims of its packaging in a manner that is authentic and clear. It strives to communicate an appeal based on trust inspired by the skill and traditions of artisans. It consists of a qualitative approach, and a tribute to the raw, natural, and organic material.

Despite the extraordinary rise of new materials and techniques, the appeal for more bygone materials and techniques still exerts a strong pull. Glass bottles, fabric bags, molded paper, tin cans, corks, and wax seals still attract consumers even though they have been gradually replaced by plastic substitutes. Still, are we to believe that the packaging methods of the past were better? In fact, they were simply adapted to the distribution method of a given time and would now be difficult to entirely recreate. Indeed, the choice of using the organic materials, despite their natural appearance can sometimes be what I like to call a "green illusion." In large amounts, materials such as wood for example, can have a negative impact on resource preservation. In some cases, traditional materials, because of their weight in comparison with plastic, could end up being a questionable choice. Heavier materials have a direct impact on transportation costs, which as a result, can increase the ecological impact. In fact, every design decision is a matter of context. What applies to a local or niche market, might not necessarily work in mass distribution, and vice versa. Nonetheless, bags, boxes and cases, and all those extra, special components that enhance the value of the product should have another use even after the product has been consumed.

The Down to Earth approach to packaging will most probably enhance the product, but with a focus on the overall experience of the user. Ultimately, it is a manifestation of the producer's consideration for their product, and for the user. In form and function, the package takes part in a ritual that is specific to the product, and that revives practices and traditions of yesteryear. While some might see a form of rejection of modernity, others could see a return to a quest for authenticity in our consumption. The intent is not to regress back to into another time, but rather to influence the market with humanity and caring reminiscent of the olden days.

ENVISION : DESIGN

1 Product: 1-2-Grow!, Client: Tropica, Distribution: Denmark, 2009

GOODMORNING TECHNOLOGY

2 Product: Scanwood, Distribution: worldwide, 2009/2010

The use of a few simple design elements—grass, soil, and roots—illustrate the environmentally friendly production process of this wooden appliance manufacturer, giving consumers across all languages a sense of buying a product that grows straight out of the ground.

1

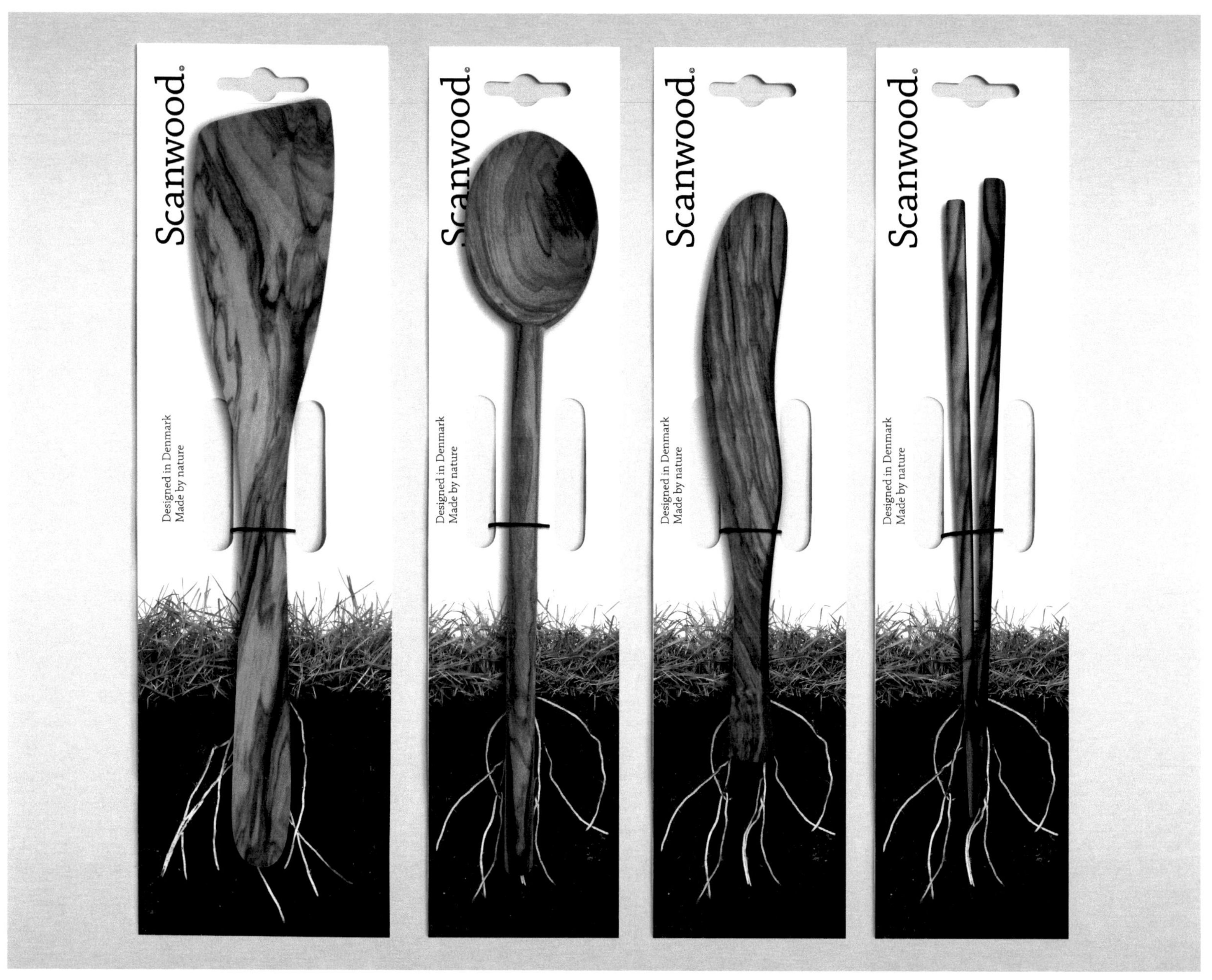

2

TREASURE

3 Product: Sustain, Client: Bentley Organic, Distribution: UK, 2011

On the packaging of these soap bars, made from sustainable palm oil, a texture-based illustration style is used along with creative copy not only to stand out on the shelf, but also to inform consumers of the negative impact by non-sustainable palm oil production on the natural habitats of endangered animals.

EYAL BAUMERT

4 Client: Holy Cacao Crafted Chocolate, Distribution: Israel, USA, Europe, 2010, Material: woodfree paper

3

4

1

THE PEOPLE'S SUPERMARKET
For the people, by the people
LAMB'S CONDUIT ST. LONDON
We're opening
THE PEOPLE'S Kitchen
THE PEOPLE'S White wine
THE PEOPLE'S Red wine
BRAND GUIDELINES
YOUR SUPERMARKET YOUR BRAND

Chantarelles
£15.00/kg.

THE PEOPLE'S Red wine
THE PEOPLE'S White wine
LAMB'S CONDUIT ST. LONDON
THE PEOPLE'S Cheese

THE PEOPLE'S Loaf

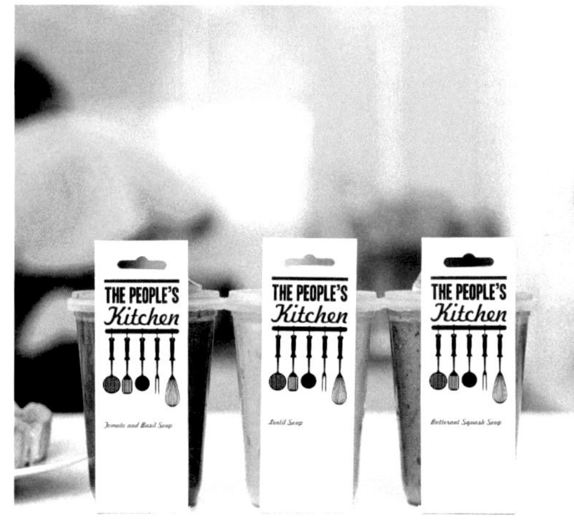
THE PEOPLE'S Kitchen
THE PEOPLE'S Kitchen
THE PEOPLE'S Kitchen

UNREAL

1 Client: The People's Supermarket, Distribution: London, UK, 2010, Material: packaging labels printed on GF Smith Colorplan Pristine White, and applied by hand using Xyron sticker machine
A simple and cost-effective packaging solution for the in-house brand of a cooperative supermarket was designed to reflect its communal, affordable, and democratic core values.

PETER GREGSON

2 Product: Granny's Secret, Client: Foodland/Granny's Secret, Distribution: Serbia, 2010, Material: paper labels with die cut, and blind print

IRVING & CO

3 Product: Rapha Performance Skincare Range, Client: Rapha, Distribution: worldwide, 2010

2

3

1

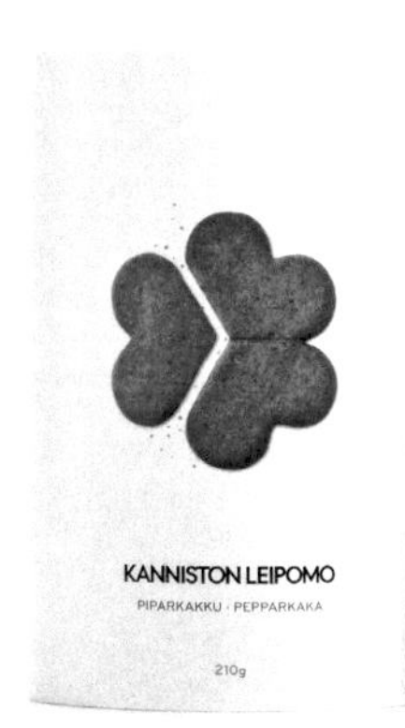
KANNISTON LEIPOMO
PIPARKAKKU · PEPPARKAKA
210g

KANNISTON LEIPOMO
PIPARKAKKU · PEPPARKAKA
210g

KANNISTON LEIPOMO
PIPARKAKKU · PEPPARKAKA
210g

KANNISTON LEIPOMO
PIPARKAKKU · PEPPARKAKA
210g

KANNISTON LEIPOMO
PIPARKAKKU · PEPPARKAKA
210g

2

zdra vo
ORGANIK
100%
KRUŠKA
750ML
ORGANIK
100%
CVEKLA
200ML
ORGANIK
100%
MALINA
380G
ORGANIK
100%
MALINA
380G
100%
CVEKLA
200ML
100%
ŠARGA-
REPA
750ML

3

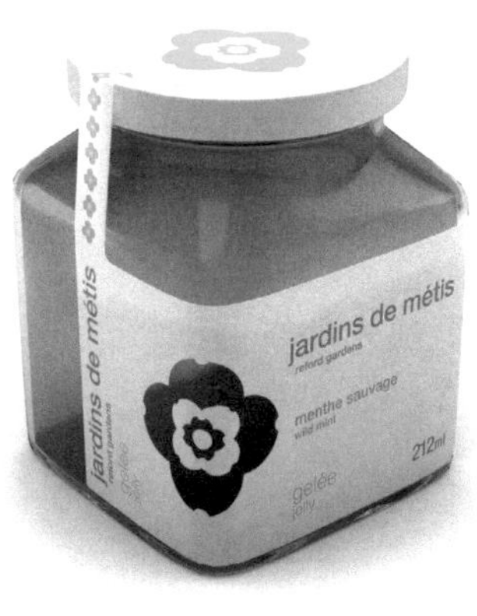
jardins de métis
menthe sauvage
212ml
gelée

jardins de métis
pommettes
gelée
212ml

jardins de métis
thé du Labrador
212ml

jardins de métis

jardins de métis

jardins de métis

4

5

PACKLAB PARTNERS

1 Product: Premium Gingerbread Packaging, Client: Kanniston Leipomo (Bakery), Distribution: Finland, 2010, Material: carton and PA/PE

For the redesign of this premium Christmas gingerbread range from Finland, subtle graphics are married with the Scandinavian tradition of breaking gingerbread into three pieces to bring good luck, while the individual heart-shaped pieces reflect a familiar yuletide motif.

PETER GREGSON

2 Product: Zdravo Organic, Client: Zdravo, Distribution: Serbia, 2010, Material: paper labels, die cut, and flexo print

A new label and glass container shape were implemented to underline the organic aspect of the juices and other food products offered by this health food company.

EMULSION

3 Product: Jelly Line Products, Client: Jardins de Métis—Reford Gardens, Distribution: Canada, 2010, Material: transparent glass jar with white metallic lid

The discreet packaging design for this line of jellies uses transparency to best showcase the product. Organic forms and counter forms serve to illustrate the superior quality ingredients that go into each gourmet product. The sweet product line features a white background while the savoury line stands out with its transparent look.

GLASFURD & WALKER DESIGN

4 Product: Dirty Apron Foods, Client: The Dirty Apron Cooking School & Delicatessen, Distribution: Canada, 2010, Material: glass containers, metal lids, adhesive label stock with die-cut 4 color process, matte varnish

HINOMOTO DESIGN - SHIGEKI KUNIMATSU

5 Product: Le Lait de la Forêt, Client: Shinrin no Bokujo Co., Distribution: Ltd., Distribution: Japan, 2010,

ATIPUS

1 Product: Fruita Blanch Handmade Jams and Preserves, Client: Fruita Blanch, Distribution: Spain, 2011

AVIGAIL BAHAT

2 Product: Cassius, student-project at Shenkar College of Engineering and Design, Distribution: Israel, 2010

STILETTO NYC

3 Product: Classic Gift Pack, Client: Rick's Picks, Distribution: USA, 2004, Material: white polypropylene label

COMPANY

4 Product: Seasonal Condiments, Client: Sheila's, Distribution: UK, 2011

A labelling system was designed for a cottage industry producing seasonal jams, marmalades, and chutneys. In order to communicate that recipes and ingredients change with the seasons, the uniform condiment jars are applied with individual labels indicating the type of product, each ingredient (defined by the ingredient color), as well as the production month, or season.

1

2

CASSIUS
Pickled Cabbage
CASSIUS
Pickled Cucumbers

Dijon Mustard
Pesto Sauce
Sun-Dried Tomatos

3

rick's picks®
bee 'n' beez™
sliced bread-and-butter pickles
hand-packed & all natural
NET WEIGHT: 15 OZ. (427g)
windy city wasabeans™
soy-wasabi green bean pickles
hand-packed & all natural
NET WEIGHT: 15 OZ. (427g)
green tomato condiment™
curried green tomato pickles
hand-packed & all natural
NET WEIGHT: 15 OZ. (427g)
phat beets™
aromatic pickled beets
hand-packed & all natural
NET WEIGHT: 15 OZ. (427g)

4

Blackcurrant
Apple
Summer
Jam
Victoria Plum
Summer
Jam
Strawberry
Gooseberry
Summer
Jam
Golden Plum
Autumn
Jam

MARKS & SPENCER
CALMING
infusion
20
TEABAGS
camomile limeflower
& lavender

MARKS & SPENCER
REVIVING
infusion
20
TEABAGS
spearmint, peppermint
& fennel

MARKS & SPENCER
DETOX
infusion
NATURALLY CAFFEINE FREE
20
TEABAGS
nettle, dandelion & milkthistle

1

2

MYRTLEFORD
the
BUTTER FACTORY
BUTTER
MILK
1L
BUTTER
MILK
300ML

BUTTER
12 x 250g

BUTTER
12 x 250g

3

STUART KOLAKOVIC

1 Product: Marks & Spencer Naturally Caffeine Free Tea Boxes, Client: Marks and Spencer, Distribution: UK, 2010, Material: recycled card stock

SWEAR WORDS

2 Product: Buttermilk, Client: The Butter Factory Myrtleford , Distribution: Australia, 2010, Material: plastic bottle, 2 color process adhesive label, 1L & 250 ml

JULIAN BAKER

3 Product: Clever/Classic Honey, Client: BHoney, 2010

STUDIO LAUCKE SIEBEIN

4 Product: Blütenhonig, Sommerhonig, Waldhonig, Client: Apifaktur, Distribution: Germany, 2010, Material: self-adhesive paper, honey glass, offset printing

4

DEPOT WPF
Product: Mlk Dairy, Client: Mlk, Distribution: Russia, 2010, Material: glass, plastic, carton

mlk
low fat milk
0.5%
organic dairy

mlk
mild
yoghurt
5.5%
organic dairy
mlk
full
cream
milk
10%
organic dairy

FRANK ALOI

1 Product: Heirloom Seed Packs, Client: The Little Veggie Patch Co., Distribution: Australia, 2011

THE CREATIVE METHOD

2 Product: Alternative, Client: The Creative Method, Distribution: Australia, 2010, Material: laser engraved on balsa wood

EDUARDO DEL FRAILE

3 Product: Tierra/ Earth, Client: Agapito Rico, Distribution: Spain, 2010, Material: bordelesa bottle

The soil-covered commemorative wine bottle recalls the direct origin of the vine and the wine's purity and aims to awaken an emotional response with a single texture.

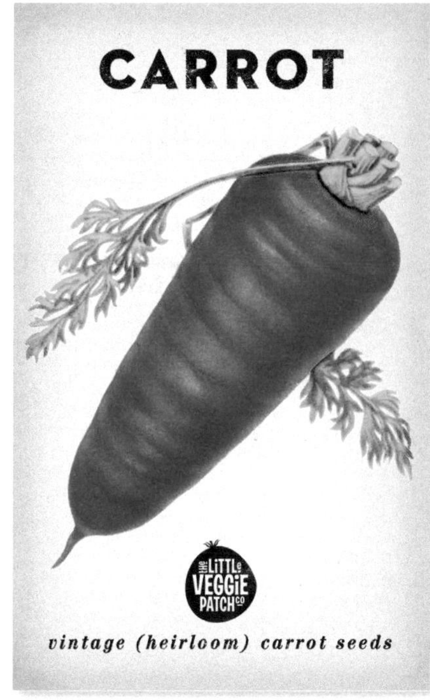

1

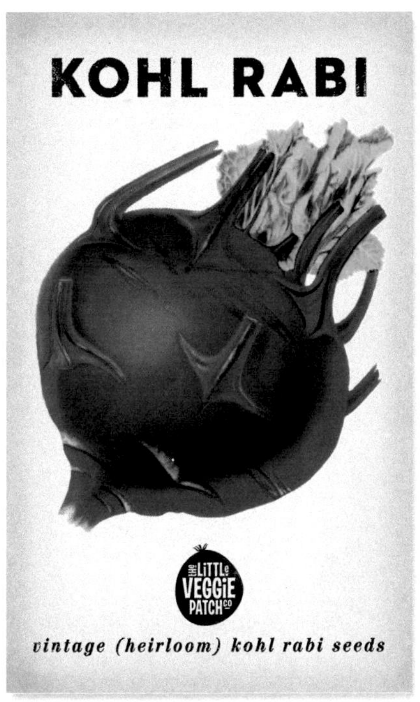

2

ALTER
NATIVE
ORGANIC
SAUVIGNON BLANC
VINTAGE
WINE OF NEW ZEALAND
12%ALC/VOL 750ML

3

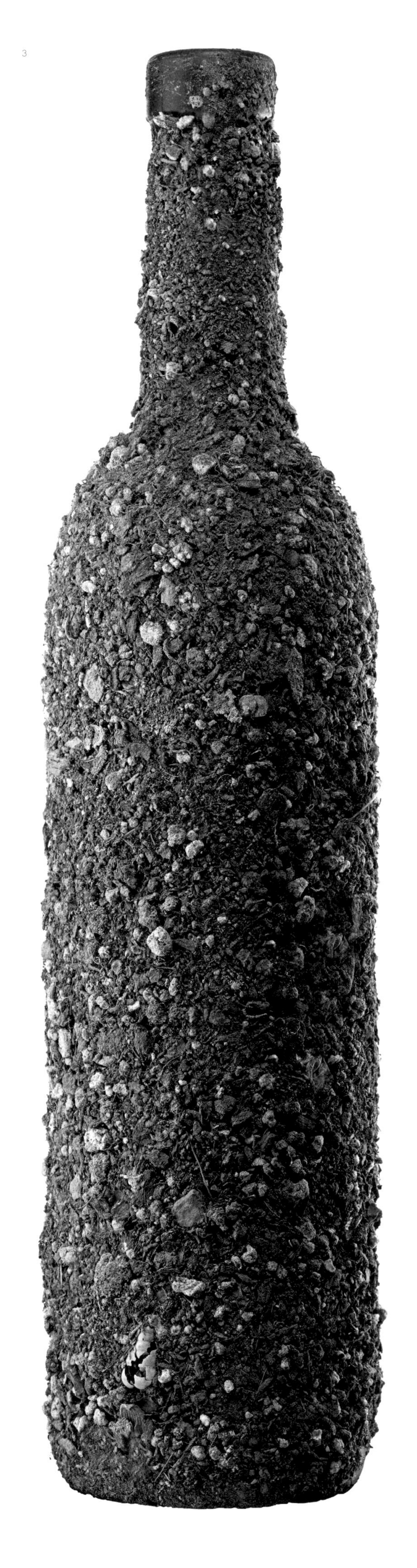

DOV KROLL

1 Product: Bara, Client: Bara Goat Dairy, Distribution: Israel, 2011

ENVISION : DESIGN

2 Product: Knuthenlund, Client: Knuthenlund Gods

1

2

ERICA CRAIG

3 Product: Kingswood, Client: Kingswood Winery, 2010, Material: constructed out of repurposed pine-beetle wood, sanded smooth, and hand painted with a wood stain

TAKU SATOH DESIGN OFFICE

4 Product: Nikka Whisky Pure Malts, Client: The Nikka Whisky Distilling Co., Distribution: Ltd./Pure Malts, Distribution: Japan, 1984

3

4

OWEN&STORK

1 Product: Atelier N°001 Owen&Stork for Portland General Store, Client: Portland General Store, Distribution: USA, Canada, UK, Australia, 2011, Material: wood, leather, glass, recycled paper, elastic shock cord, jute, cast acrylic

COLIN GARVEN

2 Product: Cask Strength Whisky, Client: Legacy , Distribution: USA, 2011

HERE DESIGN

3 Product: The Balvenie Ambassadors Case, Client: The Balvenie, Distribution: worldwide, 2009

DEPLOY UNITED

4 Product: Manifique Sheik Shaving for the Modern Man, Client: Manifique, Distribution: USA, 2010

1

2

3

THE BALVENIE
Miniature Still
THE BALVENIE
Malted Barley
THE BALVENIE
Staves

4

1

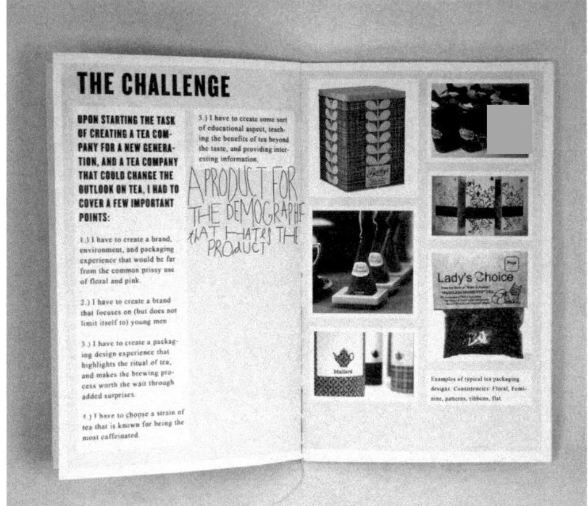

MATTHEW ALEXANDER MANOS

1 Product: Black Tea Gift Set, Client: Black Tea, Distribution: USA, 2009

The tea brand and its packaging system catering to the anarchistic demographic includes standard punk accoutrements such as safety pins, a backpack, Ziploc bags, and paper envelopes, hand-printed with a carved linoleum block. An accompanying fanzine offers a history of tea as well as several brewing processes.

MARK JOHNSON

2 Product: Understory Chocolatiers, student-project, 2009, Material: Neenah classic columns paper, neenah oxford paper, sealing wax, foil

Part of a comprehensive branding project conceived for a fictitious company based in the Central American rainforest, the elegant packaging for Understory Chocolatiers positions the company as one that cares about every step of the chocolate-making process, from bean to bar.

GRAFIKART

Edward Pearson

3 Product: Zeitun, Client: Pino Azul S.A., Distribution: Chile, 2009/2010, Material: corked glass bottle sealed with wax and paper

Packaging for this olive oil was designed to position it as a gourmet product continuing the legacy of Chilean olive oil. Screen-printed vintage bottles are corked, and traditionally sealed with embossed wax and paper, while the austere and neutral graphics emphasize the pureness of the production process, and the superior quality of the product.

2

UNDERSTORY
Chocolatiers
MILK CHOCOLATE
HOT COCOA MIX
UNDERSTORY
Chocolatiers
DARK CHOCOLATE
CHOCOLATE MILK
UNDERSTORY
Chocolatiers
DARK CHOCOLATE
CHOCOLATE MILK
UNDERSTORY
Chocolatiers
DARK CHOCOLATE
CHOCOLATE BARK

3

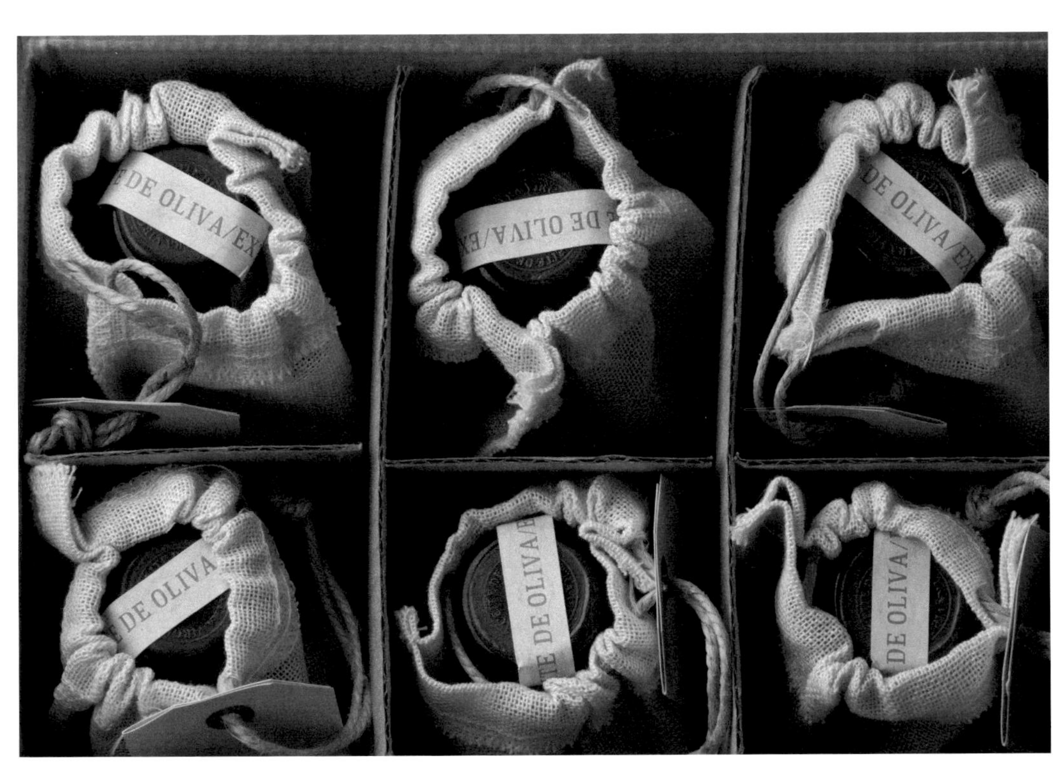
DE OLIVA/EX
DE OLIVA/EX
DE OLIVA/EX

ZEITUN
½ lt.

CO DESIGN AGENTS & THE GREAT SOCIETY

1 Product: PureProject Launch Kit, Client: Brooks Running, Distribution: worldwide, 2011, Material: bamboo, cardboard, steel

As a dramatic departure from this running shoe company's other products, its PureProject line required an equally dramatic unveiling. Fashioned from bamboo and with lids etched with the project logo, the promotional launch kit embodies the line's sustainable ethos in which nothing goes to waste. Each of the four raindrop-shaped canisters echoing the logo's shape holds a women's and men's sample of the four different shoe models. A manual for creative upcycling of the kit is also included.

SRULI RECHT

2 Product: R¿ng, Client: Sruli Recht, Distribution: Iceland, 2009, Material: 215 pieces of card, glued and painted

3 Product: Carbon Dater, Client: In-house Release, Distribution: Iceland, 2009, Material: 247 pieces of laser cut card, glue, brazing rod hinge, and ink

The cases crafted by this designer to present his own limited edition products are just as much works of art as the products that they contain. For the Carbon Dater, a black diamond-tipped carbon pen for writing directly into glass, 247 pieces of corrugated cardboard are glued together into a hinged box, and then hand painted in thematic black. The R¿ng is safeguarded on a cardboard stack that is housed inside a white-washed jewelry box with threaded lid, made from 215 cardboard pieces. The inside mount of the box holds a rough-cut diamond and white gold ring, along with two interchangeable diamonds set into propeller claws.

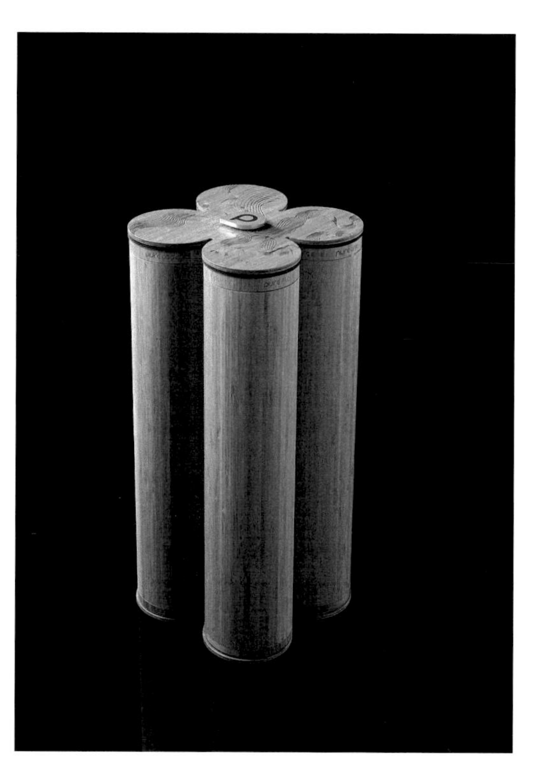

1

2

3

1

2

MATTE

1 Product: Rituals Perfume, Client: Rituals, Distribution: Netherlands, Spain, UK, Germany Portugal, Belgium, Croatia, Czech Republic, Denmark, Ireland, Lithuania, Italy, Sweden, 2010, Material: paper, wood, glass, metal, various sizes

ALEXA LIXFELD

2 Product: FRA EDP 001, Material: FRA EDP 002, FRA EDP 003, Client: Alexa Lixfeld Design GmbH, Distribution: worldwide, 2009, Material: Glass, concrete

3

4

5

6

MARLEY STELLMANN

3 Product: Barter Shopping Bag, Client: Barter, Artistic Co-op, 2009, Material: organic cotton, hemp, reclaimed paint-brushes, thermochromic paint

AESTHETIC MOVEMENT

4 Product: Shoe Brush and Box, Client: Izola, Distribution: USA, 2011, Material: one color offset printed cardboard box

5 Product: A Years Supply Of Toothbrushes, Client: Izola, Distribution: USA, 2010, Material: cardboard, plastic, and bamboo

6 Client: Further Products, Distribution: USA, 2010

1

CHRIS HANZ FOR &OR

1 Product: S/S 11 Product Family, Client: Orontas, Distribution: Canada, USA, Australia, Europe, 2011

COLIN SPOELMAN AND DAVID HASKELL

2 Product: Moonshine and Bourbon Whiskey, Client: Kings County Distillery, Distribution: USA, 2010, Material: glass flask bottles, aluminum caps, typed paper ribbon; shrink-sealed

ÉVA VALICSEK

3 Product: Egg Box, self-initiated, 2010

2

3

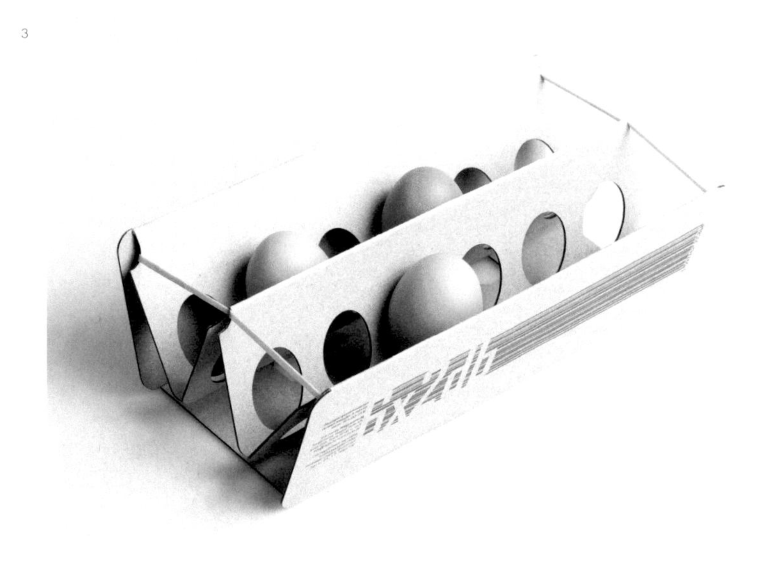

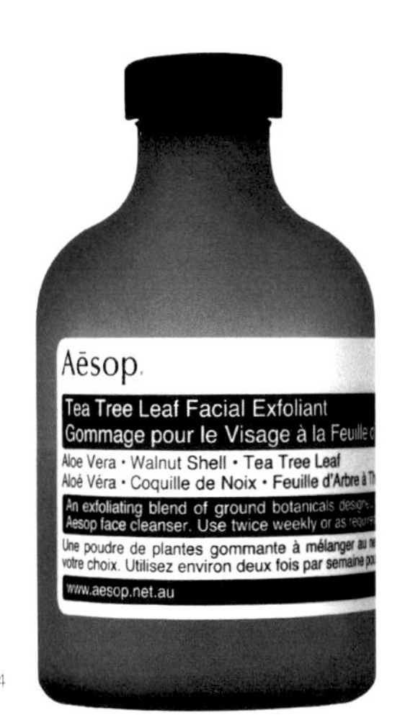

4

AESOP

4 Product: Parsley Seed Range, Client: Aesop, Distribution: worldwide, 2001-2009, Material: glass, aluminium

Sparse, clear lines juxtaposed with conceptual flair characterize the identity developed for this skin care brand. While the products themselves are packaged in simple apothecary style with rigorous homogeneity, the context of their presentation in the brand's boutiques varies from store to store. The design of each store echoes the minimalist packaging approach by focusing on a particular material, such as wood, cardboard, the amber glass bottles used to contain the products or even coconut husk string, inspired by the twine used to wrap the company's gift boxes. The power of repetition is harnessed to showcase the products, whose packaging seems all the more imposing through the sheer numbers of their occurrence.

MARCH STUDIO

5 Client: Aesop, 2010

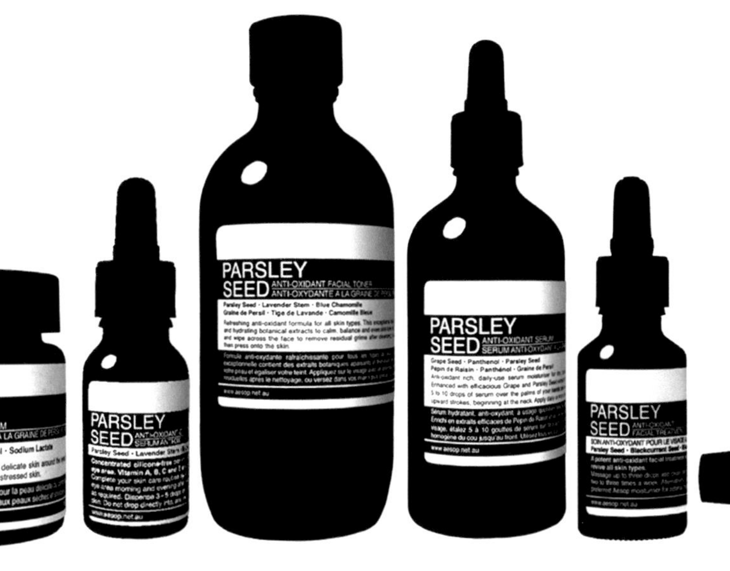

5

MARCH STUDIO
Client: Aesop, 2010

CAVALIER
SHOE POLISH
PROTECTS & SHINES
BROWN
CAVALIER
C
CAVALIER
AFTER SHAVE
ANTI-IRRITATION BALM
CAVALIER
CAVALIER
C
CAVALIER
ASPIRIN
ACETYLSALICYLIC ACID

CHAPTER 4

GOOD OLD DAYS

Until the early 1980s Canadian beer bottles had a small neck that gave them a stubby shape; it is characteristic of the time when, as a young adult, I used to hang around the local bar in my small hometown. A few years later, one of the major Canadian breweries decided to stretch the neck of its bottles, and all the other brands quickly followed suit. In just a few months, the traditional short brown bottle had disappeared from the market. Sitting on a Montreal bar terrace the other day, I was struck to find out that a small local microbrewery, for distinction purposes I guess, brought back this typical short bottle of yesteryear. Fascinated by this little anachronistic package, I told the server, "it's great to see small bottles from the 80s back on the market." The server replied: "Yeah, it's so ancient ..." A bit stunned at the word ancient, which somewhat emphasized our age difference, I politely replied: "If you don't mind I'd rather use retro when I talk about the 80s."

In a society based on consumption where brands succeed at a frenetic pace, each confirming the obsolescence of the preceding one, to some the retro style might seem archaic. However, the excitement associated with this insatiable quest for novelty in the development of new brands can also generate a certain apprehension of the consumer because of the absence of reference points. Although all historical references to the past are no guarantee of comfort, retro-style packaging tends to speak about continuity and sustainability, and is therefore reassuring to us. Products that have passed through time and trends can have some advantages over newer brands. They tell a story of a time that we generally have not experienced, and therefore it can allow a certain amount of idealization.

The Good Old Days trend is particularly noticeable in food, liquor, and body care products. This attraction to the past is often manifested by a fascination for antiquated traditions, and know-how as well as a quest for authenticity. From a graphic point of view, wood type, illuminated manuscripts, art deco typography, and handwriting all resurface in a language clearly borrowed from the past, but updated to respond to today's standards. In addition to the formal elements borrowed from the past, the retro approach also opens the door to more innocent communication strategies as opposed to today's measured and brand dictated discourses.

Each period is linked to a number of codes and visual references that have been transmitted more or less objectively by literature, film, photography, and many other media. The retro style is also the desire to identify a product with a number of commercial and social values associated with a given time. It often celebrates periods marked by a perception of commercial freedom. In reaction to the polished discourse of current marketing, the retro style would seem to allow for certain anachronistic misdemeanors to take place—such as updating 1950s-era pinups to sell chocolate. At times this style tends to favor a more popular language, and a more down-to-earth branding approach, which borrows its sense of naivety from another era. These references to the past also seems to permit a certain ironic humor as we see in the case of the *"Thickest Human Snot"* and *"Olde Fashioned Brain Jam"* by *Hoxton Street Monster Jam Supplies* or with the *"No Tell Motel Soap"* by *Blue Q*.

In its reference to the past, the look of the Good Old Days may at times convey a certain apprehension about our present, or indeed our future. In this way it embodies a certain resistance to a global and uniform culture. However, this quest for authenticity and trust in the retro style brand can also be a trap that leads to the overuse of retro codes. Many contemporary products do not necessarily have a history behind them, and so may not be well suited for the retro style. Authenticity itself is surely questioned when a new product without a past invents an artificial history for itself.

1

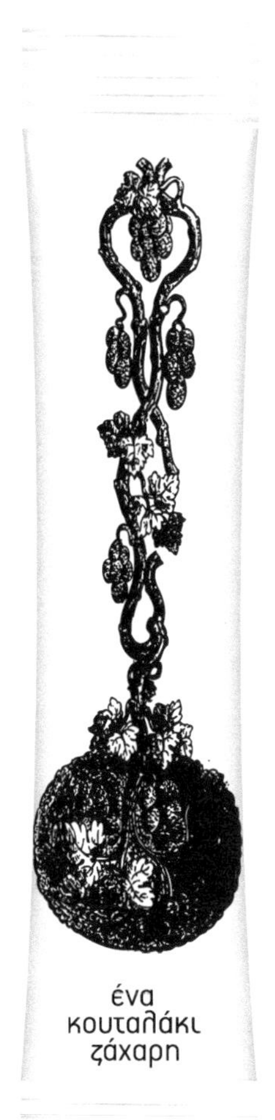

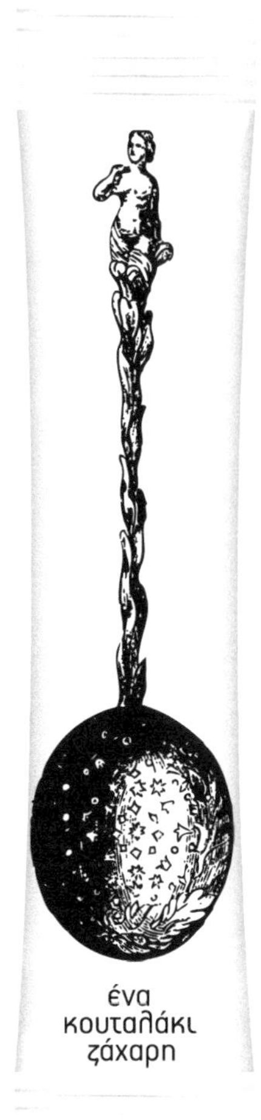

MOUSE GRAPHICS

1 Product: Sugar Sticks , Client: Sugarillos S.A., Distribution: Greece, 2009, Material: mono-dose sugar sachets

The packaging reflects its contents in these mono-dose sugar sachets. Each packet, which contains the equivalent of a teaspoonful of sugar, depicts a classic teaspoon in floral or ornamental design.

APARTMENT ONE

2 Product: Shopping Bags, Client: Moomah, Distribution: USA, 2009

2

3

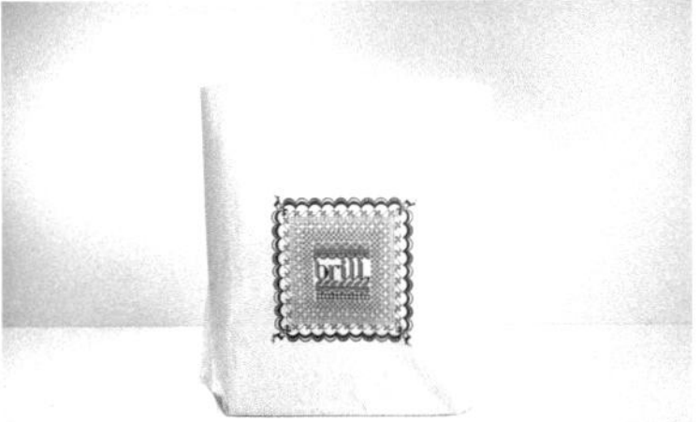

COUPLE

3 Product: Brill, Client: Simply Bread, Distribution: Singapore, 2008, Material: various packaging mainly produced in paper and plastic

A no-frills packaging solution was devised for this urban take-out food business. To match the store logo, a customized roll of tape was produced to label the complete range of products with varying shapes and sizes, ranging from plastic bag, and napkin to a single pouch for eating on the run.

CH C LATE RESEARCH FACILITY
Autumn Winter 09: Triple Chocolate
CH C LATE RESEARCH FACILITY
Autumn Winter 09: Stout
CH C LATE RESEARCH FACILITY
Autumn Winter 09: Maple Syrup
CH C LATE RESEARCH FACILITY
Autumn Winter 09: Horlicks
CH C LATE RESEARCH FACILITY
Autumn Winter 09: Cookies And Cream
CH C LATE RESEARCH FACILITY
Autumn Winter 09: White Truffle
CH C LATE RESEARCH FACILITY
Autumn Winter 09: Triple Chocolate
CH C LATE RESEARCH FACILITY
Autumn Winter 09: Apple Pie
CH C LATE RESEARCH FACILITY
Autumn Winter 09: Maple Syrup
CH C LATE RESEARCH FACILITY
Autumn Winter 09: Caramel With Sea Salt
CH C LATE RESEARCH FACILITY
Autumn Winter 09: White Truffle
CH C LATE RESEARCH FACILITY

1

2

ASYLUM

1 Product: Chocolate Research Facility—Autumn/Winter 2009, Client: Chocolate Research Facility, 2009

HATCH DESIGN

2 Product: Cowgirl Skincare, Client: Sanitas Skincare, 2010

R DESIGN

3 Product: Christmas 2010 – Noel and Windsor Forest, Client: Crabtree & Evelyn, Distribution: worldwide, 2010

MARKO GERECI

4 Product: Aroma Herbal Tea - Camomile, Yarrow, Green Tea, Client: Aroma, 2010, Material: cardboard

3

4

LOUISE FILI

1. Product: Irving Farm Coffee, Client: Irving Farm Coffee Company, Distribution: USA, 2010
2. Product: Il Conte, Client: Polaner Selections, Distribution: USA, 2009

SANDSTROM PARTNERS

3. Product: Hot Cocoa Tins, Client: Moonstruck Chocolate, Distribution: USA, 2010
4. Product: Moonstruck Classic Bar Line, Client: Moonstruck Chocolate, Distribution: USA and Asia, 2010

1

2

3

4

5

HELVETICA INC

5 Client: Kracie Home Products Sales, Distribution: Ltd., Distribution: Japan, 2010

LOUISE FILI

6 Product: SVA Senior Library, Client: School of Visual Arts, Distribution: USA, 2011

6

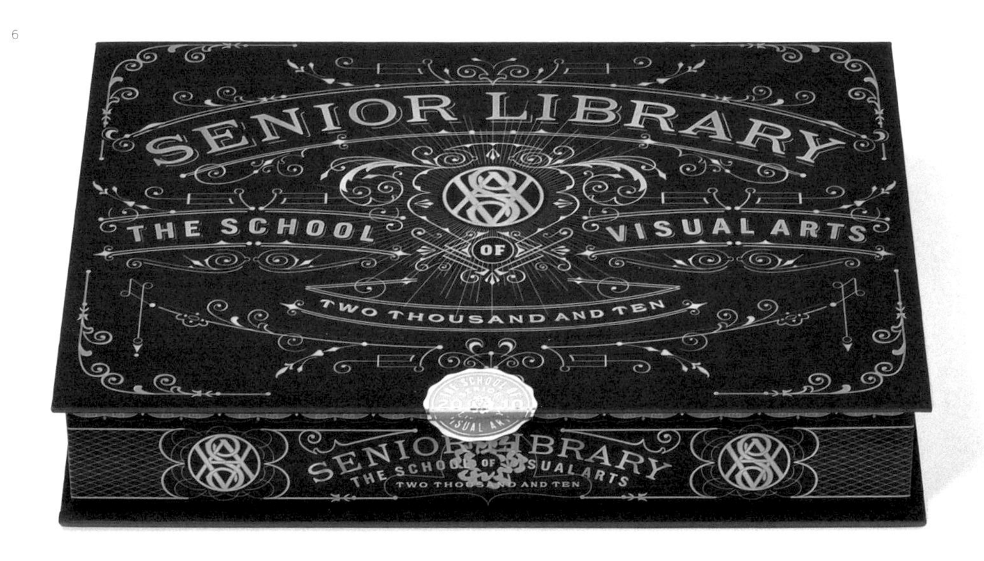

1

JONES KNOWLES RITCHIE

1 Product: Penhaligon's Christmas Boxes, Client: Penhaligon's, worldwide, 2010

JONATHAN ADLER

2 Products: Hazard Coasters, Jet Set Coasters, Hashish Candle, Elephants Salt and Pepper, Whale Salt and Pepper

2

3

4

MICHAELA BRUNNER - MEDEA DESIGN

3 Client: Péclard zurich, Distribution: Switzerland, 2010

SMITH & MILTON

4 Product: Fine Food Range, Client: Crabtree & Evelyn, Distribution: UK, Hong Kong, 2010

IRVING & CO
Julian Roberts
1 Product: Porcini Risotto Box, Client: Carluccio's, Distribution: UK, 2009

DESIGNERSJOURNEY
2 Product: Bang Chau Kaffeforettning, Client: Bang Chau Kaffeforettning, 2010

BEN BAILEY
3 Product: Dark Bar (sao tome single origin), Client: Curious Chocolate, Distribution: UK, 2010, Material: foil and recycled paper

IRVING & CO
Julian Roberts
4 Product: Panettone Tradizionale
5 Product: Panettone Ciocolatto
6 Product: Tortionata Bianco Natale
7 Product: Grissini al Cioccolato Fondente, Client: Carluccio's, Distribution: UK, 2010, Material: carton, high gloss spot color finish, foil blocking

The packaging for this range of classic Christmas delicacies produced by a London-based Italian restaurant and café chain uses a sparse, clean, and modern typographic approach with glossy paper finishes and foil-blocked graphic elements to evoke a sense of tradition, and authenticity with contemporary flair.

1

2

3

DARK
70% COCOA
SAO TOMÉ SINGLE ORIGIN

SEA SALTED CARAMEL
DARK CHOCOLATE
TRUFFLES
CURIOUS CHOCOLATE

4

Carluccio's
PA
ET
TONE
TRADIZIONALE

5

PA
ET
TONE

6

TORTIONATA
MADE WITH ALMONDS AND FRESHLY CHURNED BUTTER
Carluccio's
BIANCO NATALE
ITALIANO

7

GRISSINI AL
Carluccio's DARK CHOCOLATE GRISSINI
CIOCCOLATO

NICOLAS MÉNARD

1 Product: Pur(e) Phentex, student-project, 2011

This student packaging project revisits the cult knit slippers made of Phentex, a durable, synthetic yarn developed in the 1970s. Each cardboard box is color-coded to identify the size of the slippers. The chequered pattern reflects the slippers' typical knit pattern, while the red and white color scheme, bilingual copy, and the national maple leaf identifies the product as unmistakably Canadian. The boxes hook on to one another at the handle for efficient storage and display.

WALLNUTSTUDIO

2 Product: Pastry Shop Packaging, Client: Marypat, Distribution: Colombia, 2009, Material: cardboard, thread

Inspired by vintage suitcases that have traveled all over the world, the packaging designed for this pastry shop is a flexible, and fun collection of stickers, stamps, and cards that can be combined according to whim on various sized boxes, and bags while maintaining a consistent visual identity.

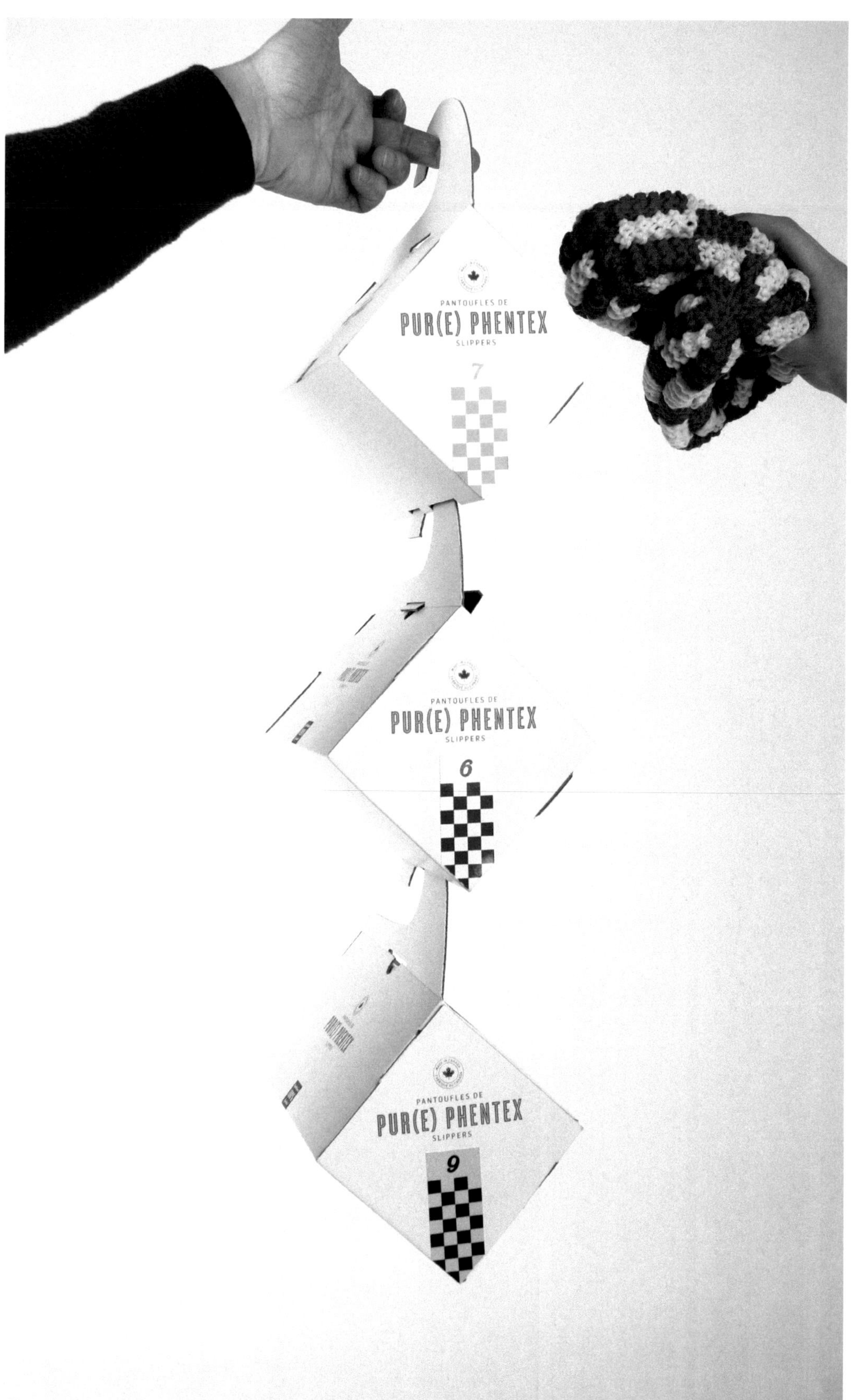

1

2

MARC L. COOPER

1 Product: COOP'S Handmade Hot Fudge, Client: COOP'S MicroCreamery, Distribution: USA, 2009

The can for this handmade hot fudge is designed to whet the appetite with the chocolate scented, brown wax that is applied by hand to the lid so that no two drips are alike. The bright yellow label, and hand-drawn aesthetic of the lettering and illustrations serve as further eye-catching elements.

ZOO STUDIO

2 Client: Rubén Álvarez, Distribution: Spain, 2010

COMMONER, INC - RICHIE STEWART

3 Product: Belgian Tripel Ale, Client: The Lower Depths, Distribution: USA, 2011, Material: screenprinted on natural cardstock

1

2

3

REGISTERED
U.S.A.
ABV: 9%
BOTTLED ON: 3.14.11
BELGIAN
TRIPEL
ALE
BREW HOUSE:
John Harvard
THE LOWER DEPTHS
SEAL OF APPROVAL
BREWED LOCAL in CAMBRIDGE MA.

EMELIE HALSTON

1 Product: Mister.e, Concept Design, 2010

SOCIEDAD ANONIMA

2 Product: Auténtico Tequila Alacrán, Client: Auténtico Tequila Alacrán, Distribution: Mexico, USA, Japan, 2010, Material: glass and matte black Soft Touch® coating

JESSIE WHIPPLE VICKERY

3 Product: Cold Brew Coffee, Client: Stumptown Coffee Roasters, Distribution: USA, 2011

4 Product: Anniversary Blend, Client: Stumptown Coffee Roasters, Distribution: USA, 2009, Material: glass Apothecary Bottle

TENFOLD COLLECTIVE

5 Product: Hare's Bride Hefewein Ale, Client: Grimm Brothers Brewhouse, Distribution: USA, 2011

MOXIE SOZO - MARLEY STELLMANN

6 Product: Concept for Good Juju, Client: Left Hand Brewery, 2009, Material: paper label on glass bottle

1

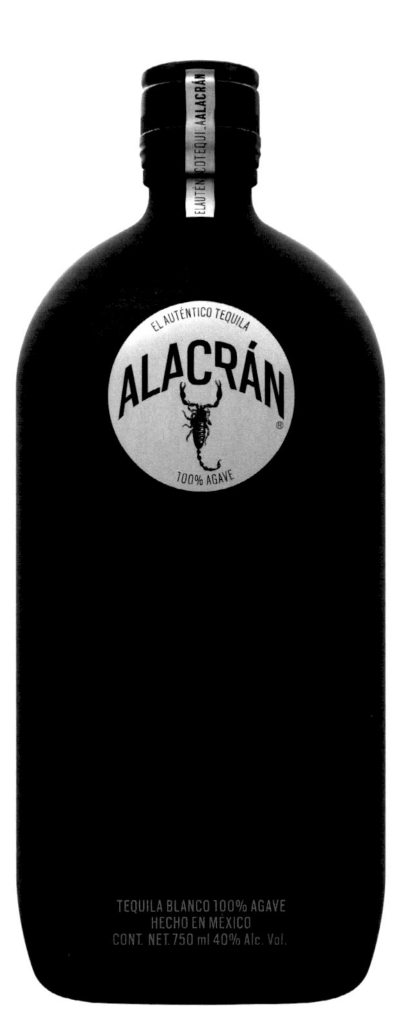

2

3

STUMPTOWN
COFFEE ROASTERS
ESTABLISHED
1999
COLD BREW
COLD BREW
COFFEE
FINEST QUALITY

4

ESTABLISHED 1999
STUMPTOWN
COFFEE ROASTERS
FINEST QUALITY
ROASTED DAILY
ANNIVERSARY BLEND

5

Fabled Series
Hare's
Bride
Hefewein Ale
1 PT. 6 FL. OZ.

6

TO SELL
NOTHING
But The Best
GOOD
JUJU
12·FL·OZ
ALE·BREWED·WITH·GINGER

MASH

1. Product: Lobo Apple Cider , Client: Lobo Apple Cider, Distribution: Australia, 2010, Material: 330ml Bottle, Cast coated labelstock
2. Product: Cashburn, Client: Burn Cottage, Distribution: New Zealand, USA, 2010, Material: 750 ml bottle, Estate 8 labelstock
The name of the first wine released by this new biodynamically farmed vineyard describes the long, and costly investment the owners made to get the vineyard up, and running as a "cashburn" project. The ornate, hand-drawn illustration on the uncoated paper Cashburn wine label depicts the mystical nature and beauty of the vineyard growing out of a pile of cash.
3. Product: Linnaea Rhizotomi, Client: Linnaea Vineyard, Distribution: Australia, USA, UK, 2010, Material: 750 ml bottle, Estate 8 labelstock
The design for the first release of this young vineyard depicts its owners as the face of the brand, illustrating their backgrounds in medical anthropology and plant biochemistry in an eye-catching, unconventional way.
4. Product: Burn Cottage Pinot Noir, Client: Burn Cottage Wines, Distribution: Australia, USA, UK, New Zealand, 2009, Material: 750 ml bottle, Estate 8 labelstock
The label art designed for this biodynamically cultivated wine is a collage representing the story *The Green Snake and the Beautiful Lily* by Johann Wolfgang von Goethe, which was a seminal text for Rudolf Steiner in his development of anthroposophy as well as the foundations of biodynamic agriculture. The fantastical pen and ink illustration evokes the 19th century symbolist art movement, and Steiner's interest in the occult.

1

2

3

4

BURN
COTTAGE
2009 PINOT NOIR
CENTRAL OTAGO
ESTATE GROWN AND BOTTLED BY:
SAUVAGE FAMILY VINEYARD
NO. 4, 7 ROGERS STREET,
CROMWELL, 9310
WWW.BURNCOTTAGE.COM
CONTAINS PRESERVATIVE 220
APPROX 8.3 STANDARD DRINKS
NEW ZEALAND WINE
ALC. 14.0% BY VOL.
750ML

CADENA Y ASOCIADOS BRANDING

Client: ADO/Brand Cielito Querido Café, Distribution: Mexico, 2010

The packaging and brand identity for this Mexican café combines vintage-style labelling and typography, and the spirit of the corner grocery store with bold chromatic applications, popular local expressions, and 20th century art, design, and architecture to create a hip, urban atmosphere that is playfully nostalgic.

MIEL
CAFÉ
COMO CAIDO DEL CIELO
CANELA
NUEZ MOSCADA
$45
$50
CACAO
LECHE
MACÉN

WILLIAMS MURRAY HAMM

1 Product: Recipease, Client: Jamie Oliver, Distribution: UK, 2008

This retail meal assembly concept is communicated through the idea of an Airfix plastic model kit. Iconic illustrations, and their various combinations create a witty language that addresses good food, and ease of cooking.

1

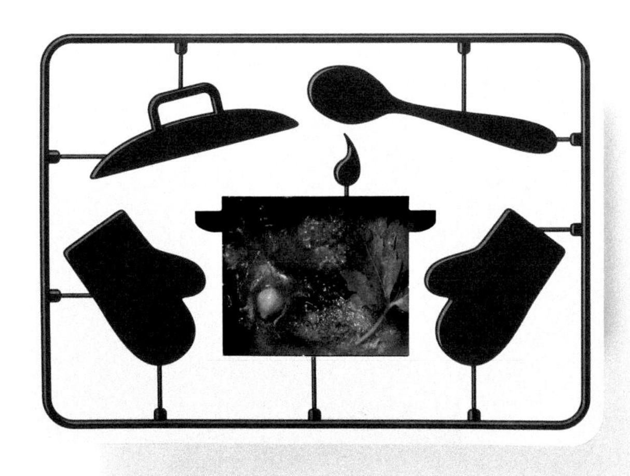

2

3

4

5

6

7

THE ROBIN SHEPHERD GROUP

Client: Bold City Brewery, Distribution: USA, 2009/2010

2 Product: 1901 Red Ale,
3 Product: Smokey Porter Tap Handle
4 Product: Abby Ale
5 Product: Fritz Hefeweizen
6 Product: Yellow Jacket Bar-B-Q Sauce, Client: Saltwater Marsh BBQ, Distribution: USA, 2010

ALVIN DIEC

7 Product: Sir Kensington's Gourmet Scooping Ketchup, Client: Sir Kensington, Distribution: USA, 2010

SOUTHERN BELLS
PROUDLY BREWED IN SAVANNAH, GEORGIA
BERRY BLUES
BLUEBERRY STOUT
NOISY NECTAR
NECTARINE WHEAT
SOUTHERN BELLS
LET THE GOOD TIMES RING
EVEN WHEN YOU'RE BAROQUE
BREWED IN
SAVANNAH
GEORGIA

B
C
D
E
USE THE FULL SET TO CREATE A SYMPHONY OF SOUND WITH YOUR FRIENDS
SOUTHERN BELLS
PROUDLY BREWED IN SAVANNAH, GEORGIA
NOISY NECTAR

SAM GENSBURG

1 Product: Noisy Nectar & Berry Blucs, Client: Southern Bells Brewery, 2011

ARCHRIVAL

2 Product: Lucky Bucket Beer, Client: Lucky Bucket Brewing, Distribution: USA, 2010

MINT DESIGN

3 Product: Monx Belgian Style Beer, Client: Monx, Distribution: USA, 2000

2

3

NIEDERMEIER DESIGN

1 Product: Candy Boxes, Client: Leatherback Printing, Distribution: USA, 2005

MARIEBELLE

2 Product: Pin-Up Girl Bars, Client: Marie-Belle New York, Distribution: USA, 2000

KYLE TEZAK

3 Product: Mindy's Hot Chocolate, Client: Mindy's Hot Chocolate, Distribution: USA, 2010, Material: cardboard

MINT DESIGN

4 Product: Voodoo Soap, No Tell Motel Soap, Client: Blue Q, Distribution: USA, 2005

CLARK MCDOWALL

5 Client: Alterra Coffe - Mars Inc., Distribution: worldwide, 2010

1

2

3

4

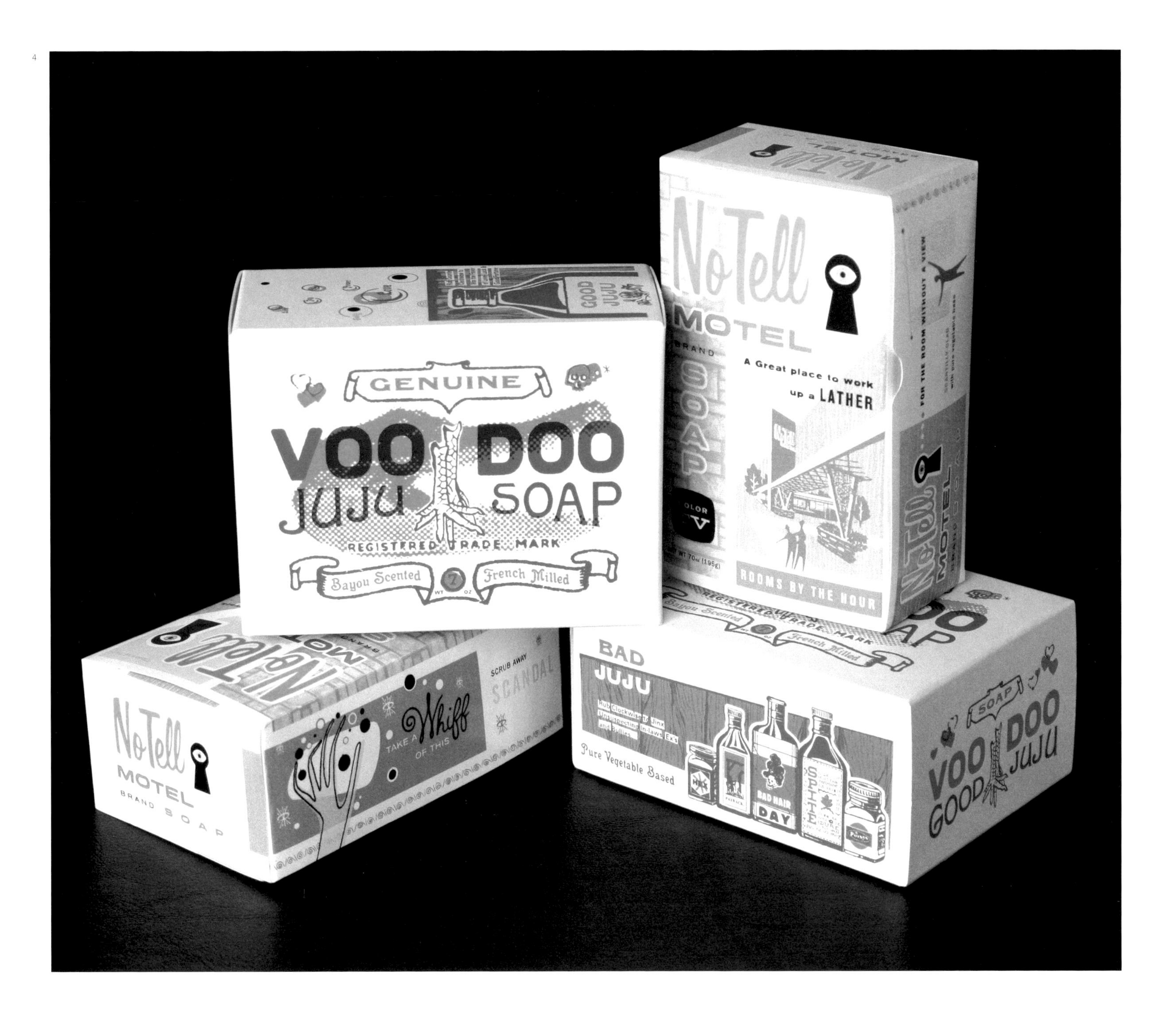
GENUINE
VOO DOO
JUJU SOAP
REGISTERED TRADE MARK
Bayou Scented
French Milled
GOOD JUJU
NoTell
MOTEL
BRAND
SOAP
A Great place to work
up a LATHER
ROOMS BY THE HOUR
FOR THE ROOM WITHOUT A VIEW
SCRUB AWAY
SCANDAL
TAKE A Whiff OF THIS
BAD
JUJU
Pure Vegetable Based
BAD HAIR
DAY
SPITE
SOAP
VOO DOO
GOOD JUJU

5

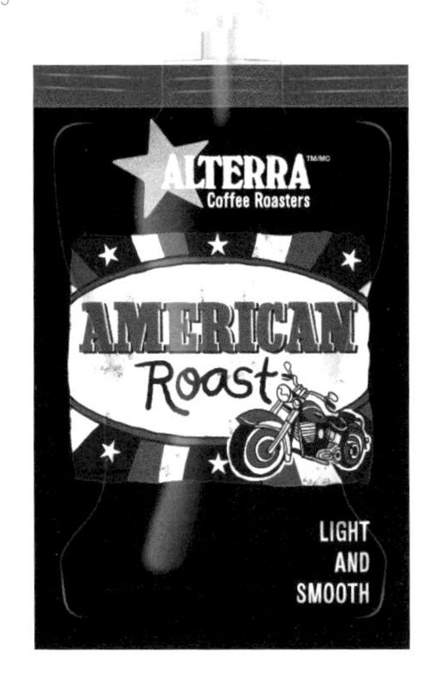
ALTERRA
Coffee Roasters
AMERICAN
Roast
LIGHT
AND
SMOOTH

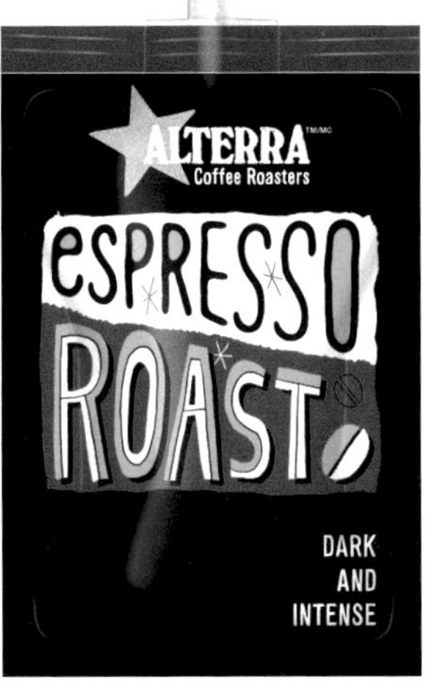
ALTERRA
Coffee Roasters
ESPRESSO
ROAST
DARK
AND
INTENSE

ALTERRA
Coffee Roasters
French
ROAST
DARK
AND
INTENSE
CERTIFIED
Over 30%
Certified Coffee

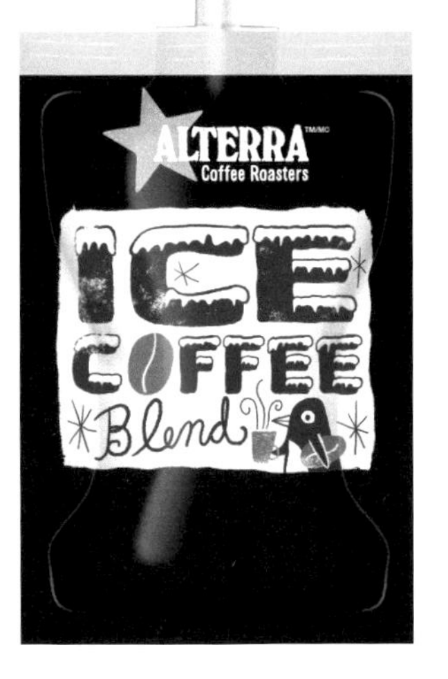
ALTERRA
Coffee Roasters
ICE
COFFEE
Blend

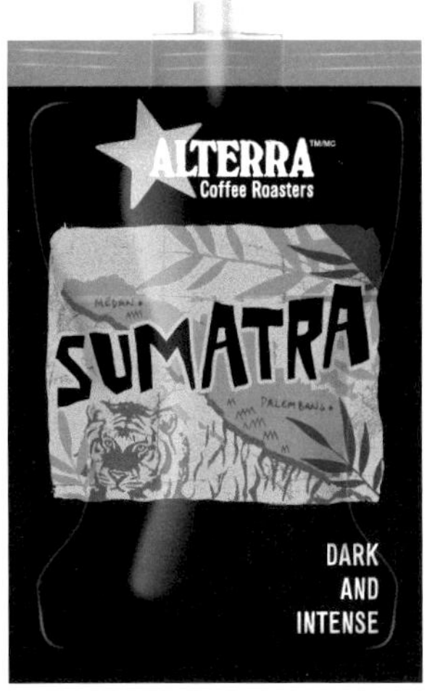
ALTERRA
Coffee Roasters
SUMATRA
DARK
AND
INTENSE

ALICE PATTULLO

1. Product: An Aid to the Amateur Sailor, Distribution: UK, 2009/2010, Material: hand screen-printed onto board
2. Product: Celebrity Smiles, 2010, Material: hand screen-printed onto board
3. Product: Mother Mutton's Aging Cream, 2010, Material: Hand screen-printed card
4. Product: Travel-safe Souvenirs, 2010, Material: screen-printed onto fabric and card
5. Product: Pattullo's Chocolate Truffles, Distribution: UK, 2010, Material: hand screen-printed onto card
6. Product: Mother Mutton's Aging Cream, 2010, Material: hand screen-printed board
7. 2010, Material: hand screen-printed onto board

1

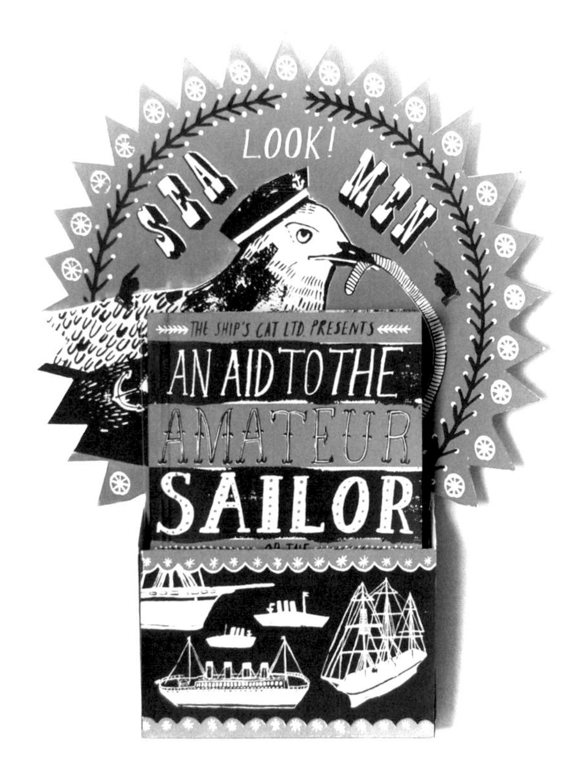

2

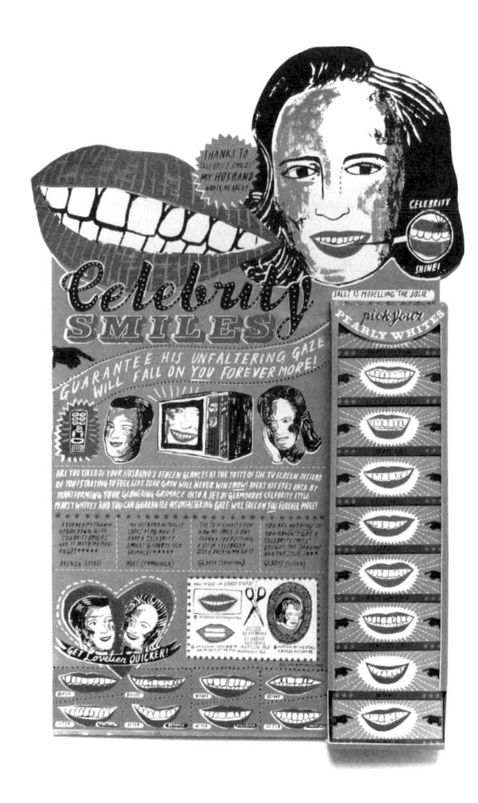

2

3

4

5

5

6

ALICE PATTULLO
PLEASE TAKE ONE
Celebrity
SMILES
POSTCARDS

BUSINESS
CARDS

HATCH DESIGN

1 Product: Annual Hatch Design Easter Egg Coloring Kit, Client: Hatch Design, 2008/2009

The packaging for this Easter egg coloring kit hearkens back to the childhood of generations past while encouraging recipients to playfully contribute to the future discourse of design.

1

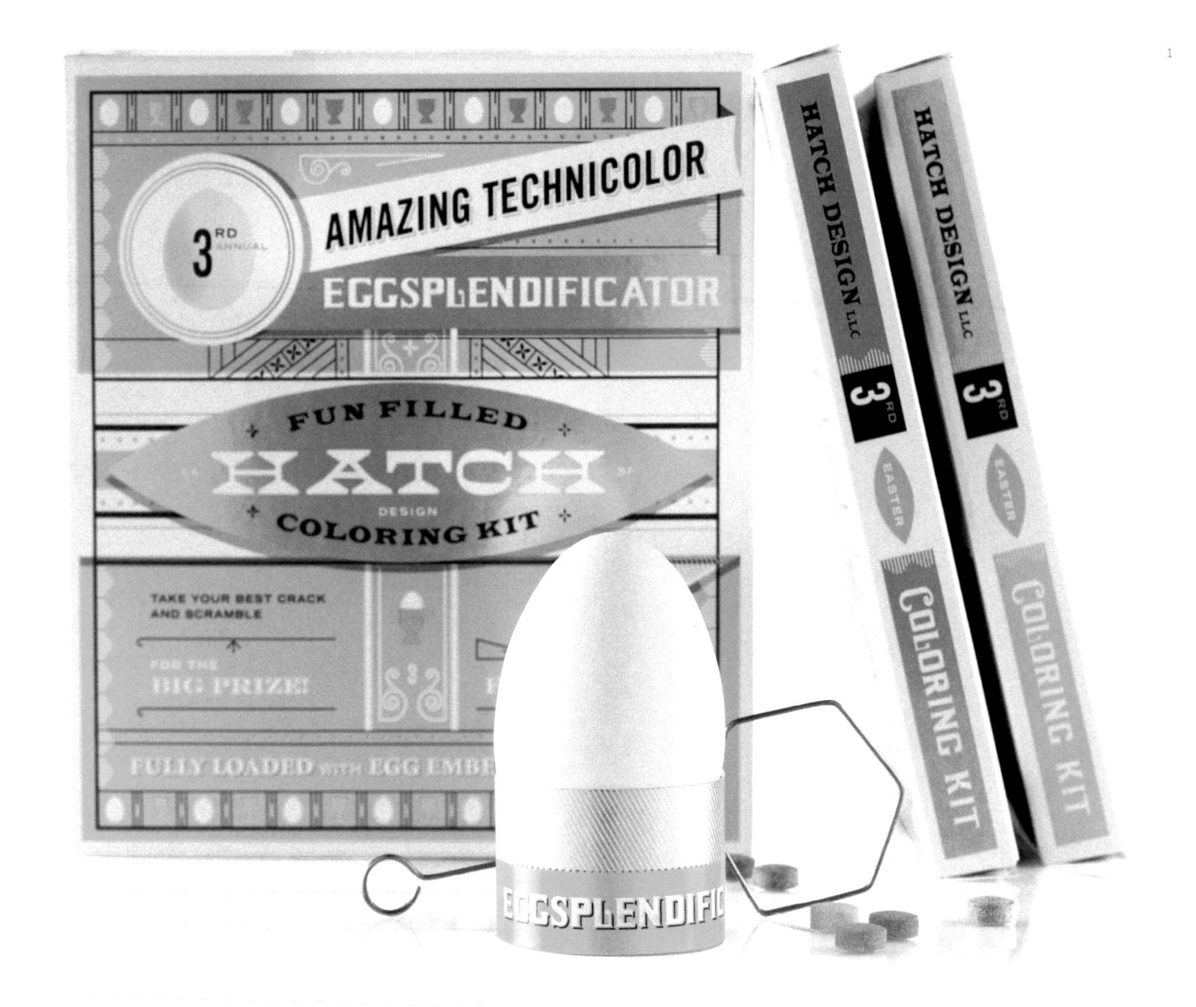

TARA BAKER

2 Product: Chatterbox Cereal, Client: Auburn University, 2011

The creation of a series of children's cereal boxes showcasing early educational themes was the goal of this student project. The whimsical, retro design and color palette of the uniquely shaped, single-serve packaging references a retro aesthetic appealing to both parents, and children.

H-57 CREATIVE STATION

1 Product: H-57 Patch, Client: H-57 Creative Station, 2010

2 Product: USB Key - Intercontinental Ballistic Design Missile, Client: H-57 Creative Station, 2010

1

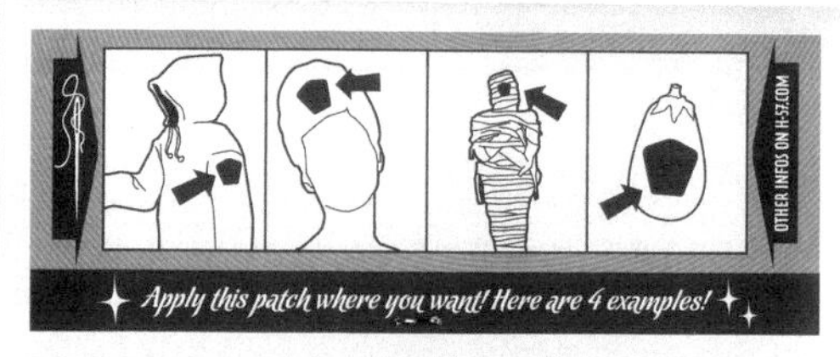

2

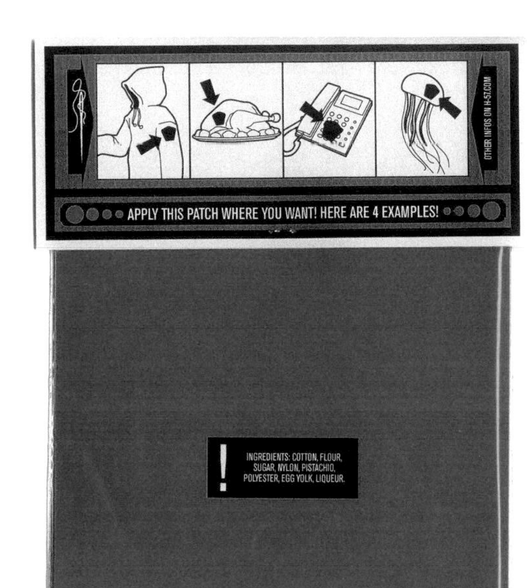

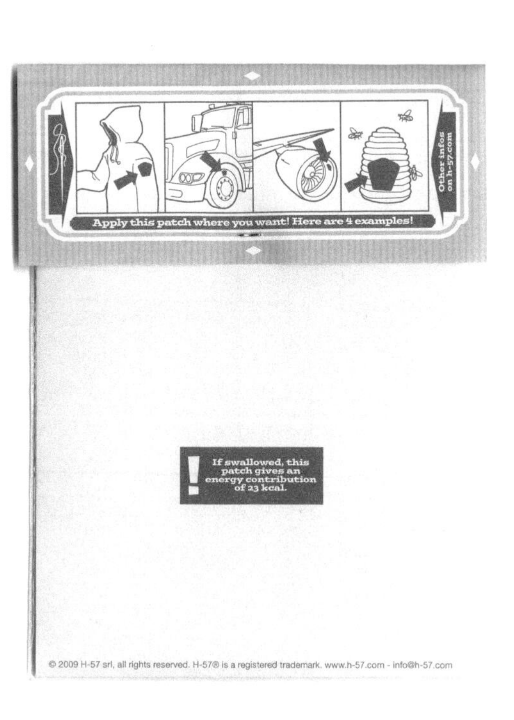

SAM GENSBURG

3 Product: Assorted Jelly Beans, Client: Curio, 2011, Material: recycled cardboard, glass, and plastic

3

ACME INDUSTRIES

1 Product: PaperBag for take-out shop, Client: Omnivore's Dilemma, Distribution: Romania, 2010, Material: kraft paper

TAYLOR C. PEMBERTON

2 Product: Cavalier Essentials, 2011

The design concept behind this line of vintage-style essentials for men bears an aura of rugged sophistication as it explains its philosophy with a question: "If Steve McQueen carried a beat-up leather duffle bag on the back of his motorcycle, what would be in it, and how would the products look?"

1

2

MERCHANT & MILLS
NEEDLE THREADER
Carolyn N.K. Denham
DRESSES COURSES PATTERNS NOTIONS

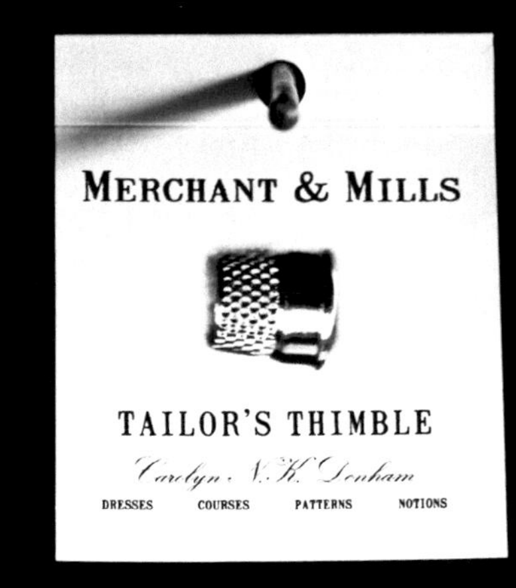

MERCHANT & MILLS
TAILOR'S THIMBLE
Carolyn N.K. Denham
DRESSES COURSES PATTERNS NOTIONS

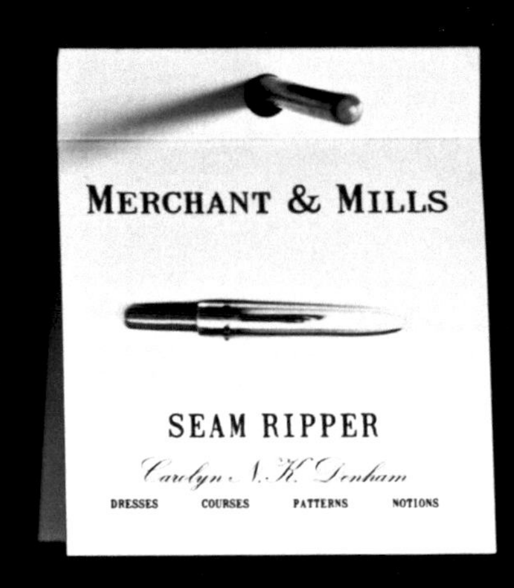

MERCHANT & MILLS
SEAM RIPPER
Carolyn N.K. Denham
DRESSES COURSES PATTERNS NOTIONS

MERCHANT & MILLS
PIN MAGNET
Carolyn N.K. Denham
DRESSES COURSES PATTERNS NOTIONS

MERCHANT & MILLS
WIDE BOW SCISSORS
Carolyn N.K. Denham
DRESSES COURSES PATTERNS NOTIONS

MERCHANT & MILLS
TAILOR'S CHALK
Carolyn N.K. Denham
DRESSES COURSES PATTERNS NOTIONS

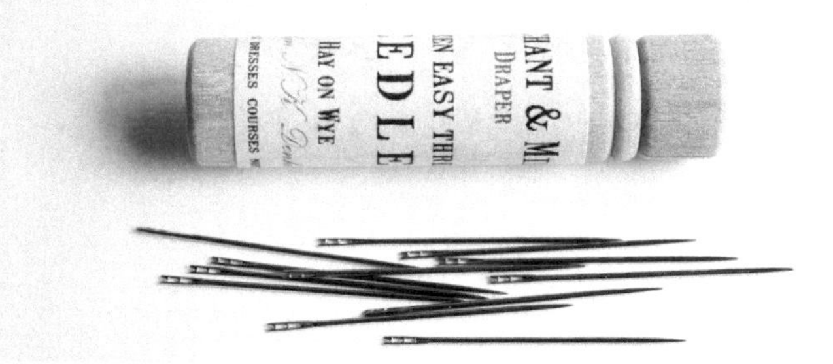

RODERICK FIELD

1 Client: Merchant & Mills, Distribution: Romania, worldwide, 2011

Simple packaging with a nod to the past, expresses the continuation of tradition in this product range. Everyday materials such as white labels, small brown envelopes, and folded paper in varying weights reference traditional packaging as well as the legacy of tailoring embodied by the brand.

2

3

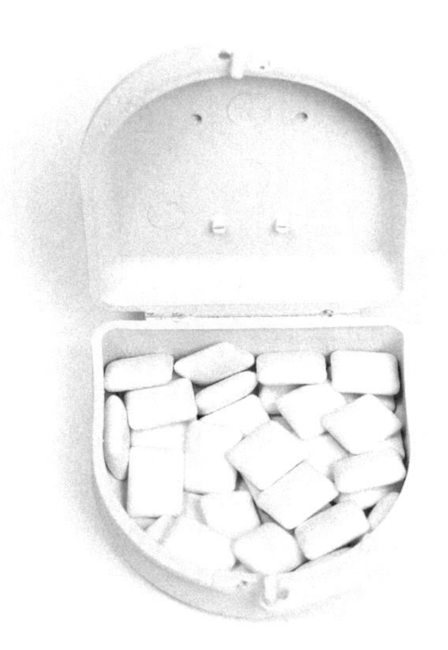

4

5

AESTHETIC MOVEMENT

2 Product: Boxed Soap Sets, Client: Izola, Distribution: USA, 2010

BRYAN CONNOR

3 Product: Pearly Whites, Client: 826 Baltimore—The Stuff of Dreams, 2010

4 Product: Instant Nightmare Rescue, Client: 826 Baltimore—The Stuff of Dreams, 2010
The playful mix of typography, witty content, and utilitarian packaging provides a tongue-in-cheek reference to the original usage of these repurposed products designed for sale at a fictional concept store. Instant Nightmare Rescue is a small package of pins. Pearly Whites are Dentine gum repackaged in a recycled teeth retainer case.

ANDREA STREETER

5 Product: Houdini Organic Soaps, Self-initiated, 2010

ILFORD

LINNA XU

Product: Ilform Film, self-initiated, Distribution: Canada, 2010, Material: paper

Designed to reintroduce, and encourage use of analogue film technologies, the packaging developed for Ilford's 120 film can be reused as a pinhole camera. Included are instructions in eight languages, and all materials required for easy reassembly of the packaging as a working camera.

TOM HAYES

1 Client: Proper BBQ, 2010

The range of barbecue essentials uses reduced and bold labelling along with natural packaging materials such as unbleached paper, cotton, and twine to recall the Wild West, and convey a sense of rugged frontiersmanship.

DAWN STEINBOCK

2 Product: Indoor/Outdoor Milk Paint, Client: Old Fashioned Milk Paint, Distribution: USA, 2010

The packaging designed for this line of milk paint reflects its traditional use on farms and antique furniture. Old wood-style typefaces are featured on the typography driven labelling, encouraging attention to flow first from the brilliant colors of the paint powders visible through the glass jars, and then to the information provided on the labels.

ARMIN BUEHLER

with Graeme Offord

3 Product: Holtz Tabak Hausmischung, Client: Holtz Tabak und Pfeife, Distribution: Germany, 2011

TRULY DEEPLY

4 Client: The Entrepreneurs' Organization, Distribution: Australia, 2011, Material: cardboard, chocolate, and paper
Recalling in style the golden ticket featured in the 1970s film Charlie and the Chocolate Factory, this Golden Ticket package, featuring a chocolate egg nestled in a cardboard container, was designed to promote an annual Easter egg hunt. Located on a small island in the center of Melbourne, the event called for a package that would contain all the mystery, and magic of its surrounding.

WE MADE THIS

5 Product: The Collywobbles, Client: Hoxton Street Monster Supplies, Distribution: UK, 2010, Material: lever lid tin, high, paper label
Hoxton Street Monster Supplies is the fantastical shopfront and identity of the Ministry of Stories, a children's writing center. The shopfront, along with the packaging of the products sold to raise money for the non-profit workshop, are designed to fire the imaginations of the participating children—and potential supporters.

5

WE MADE THIS
Product: Hoxton Street Monster Supplies,
Distribution: UK, 2010

T TIGHTENER
NECK-BOLT TIGHTENE

A VAGUE SENSE OF UNEASE

VILEST BILE
FRESH FARTS
FLAYED SKIN

Hoxton Street Monster Supplies
ZOMBIE FRESH MINTS
Suffering from hideous halitosis after gorging on humans all night? Banish brain-breath with these startlingly strong mints.
Sold Under License From The Ministry of Stories · www.ministryofstories.org

STOP
THE WATER
WHILE
USING ME!
ALL NATURAL
ORANGE WILD HERBS
SHOWER GEL

CHAPTER 5

GAME CHANGER

If I had to choose one packaging design to represent the future of packaging, I would no doubt choose the fortune cookie. It alone stands for all the virtues of a perfect package. It is both packaging and product at one and the same time. Once eaten, it leaves no residue behind other than a small message of hope. It provides a single serving, but its footprint is close to nothing. It offers a playful ritual, an experience that plays host to all the emotions linked with the pleasure of unpacking and the unfolding of a story in which we play a role. Furthermore, the fortune cookie delivers nothing less than the future itself. Quite a claim indeed! But a claim that may very well come true—and if you happen to be pleased with your fortune you very well may hold on to this delicately printed token for years to come.

Of all the issues facing the packaging industry, environmental issues remain the most relevant. The globalization of markets, the development of emerging economies, and the proliferation of packaging formats are putting increasing pressure on the price of raw materials and energy. A fundamental question arises: How to reconcile this growing increase in demand without irreversibly affecting the planet's resources and without increasing the impact of industry on the environment? Although the impact of packaging represents a small percentage of the entire global industrial footprint, it nevertheless remains one of the most visible culprits. In all honesty there is no such thing as a 100% green or environmentally friendly package. Any production process, even the production of a fortune cookie, has an ecological footprint. The challenge is to optimize the processes, reduce transportation as well as reduce both production and consumption of packaging volume.

Increasingly conscious of their image, but also of their profitability, companies are realizing that ecological thinking can be not only ideological but a cost effective choice as well. However, if the principles of ecodesign seem relatively simple, their implementation is often quite complex requiring intervention at every stage of the packaging life cycle. Care must be taken to avoid green washing, which is often nothing more than a ploy to move the environmental load upstream, or downstream of the production chain.

The environmental question has substantially changed the designer's work and forced them to integrate this paradigm into practice. As a professor of packaging design at the University of Quebec in Montréal (UQAM), I push my students to embrace a holistic approach to design and to maintain a critical view of the discipline. The goal is simple: do more with less. Not only must we protect and value materials but we must also limit the energies required for their transformation.

What might appear as a loss of creative freedom is in fact a unique challenge for research and innovation in design. A new approach inevitably leads to a new aesthetic—one based on intelligence and restraint rather than on excess and luxury. The era of carefree consumption has come to an end. Gold and silver have given way to more humble and raw materials. Chic has been redefined.

A basic yet fundamental question I ask my students is also a simple one: "Do we need another package?" Looking at what is on the market, it seems this basic question is not asked often enough. For example, do software products still need to be contained within such large packages? Given the ease with which we can now download applications and PDFs, can we not imagine the complete elimination of software packaging? And if visibility on the shelf is still required, are there not ways to reduce it to its simplest form? The role of the designer need not always involve the creation of a new object. Objects can also be reduced or even eliminated. It all has to do with our perception of the designer's mission. If designers are simply performers, then they will simply conceive a new box as requested, but if a designer acts more as a consultant, they will weigh all the problems and may very well come back with an optimal solution to suit the producer, the consumer, and the environment. Design has the privilege of being upstream from production. It is therefore located at a crucial and decisive position where we can design and implement more efficient solutions that reflect the basic principles of sustainability while balancing economic, ecological, and social values.

I often tell my students to let their ideas live. Sometimes great answers lie along the borders of utopia. An unusual creation can sometimes be a potential solution to another problem. The best guarantee for innovation is to expand beyond the frontiers of the known so that free exploration and creativity can find expression. While in today's context some concepts may seem unrealistic, they may be realizable in a not so distant future. After all, research and innovation require—at least to some extent—a rejection of the status quo.

H-57 CREATIVE STATION

1 Product: Re-Pack Project, Client: H-57 Creative Station, Distribution: Italy, 2010
The Re-Pack Project was developed to encourage consumers to re-use old packaging. Instructions on how to turn a cardboard box inside out for further use were provided, along with a red sticker encouraging the next recipient of the box to do the same.

OFFICE

2 Product: The eBay Box, Client: eBay, Distribution: USA, 2010
In order to make shipping a little greener, a new set of boxes was developed for eBay made with 100 percent recycled content, printed with water-based inks, and designed to require minimal tape. Friendly illustrations and copy encourage the re-use of the box, listing potential benefits to the planet and tips for greener packing. A little bird asks, "Where to next?" prompting the box user to write a note, so the next person to receive it can see just how far it has come.

HAPPY CREATIVE SERVICES (INDIA) PVT LTD

3 Product: Never Wasted Bag, Client: Lee, Distribution: India, 2010, Material: ITC board

YVES BÉHAR | FUSEPROJECT

4 Product: Clever Little Bag, Client: Puma
The Clever Little Bag was developed as a green alternative to the traditional shoebox. Inside, a die-cut cardboard sheet is folded to create a four-wall structure that allows for secure stacking. The bag itself, made of non-woven, heat-stitched polypropylene not only protects the shoes from dust and dirt at the warehouse, and during shipping, but also replaces the shopping bag, and can later be used by the owner for shoe storage during travel. Both the cardboard and the bag are recyclable.

1

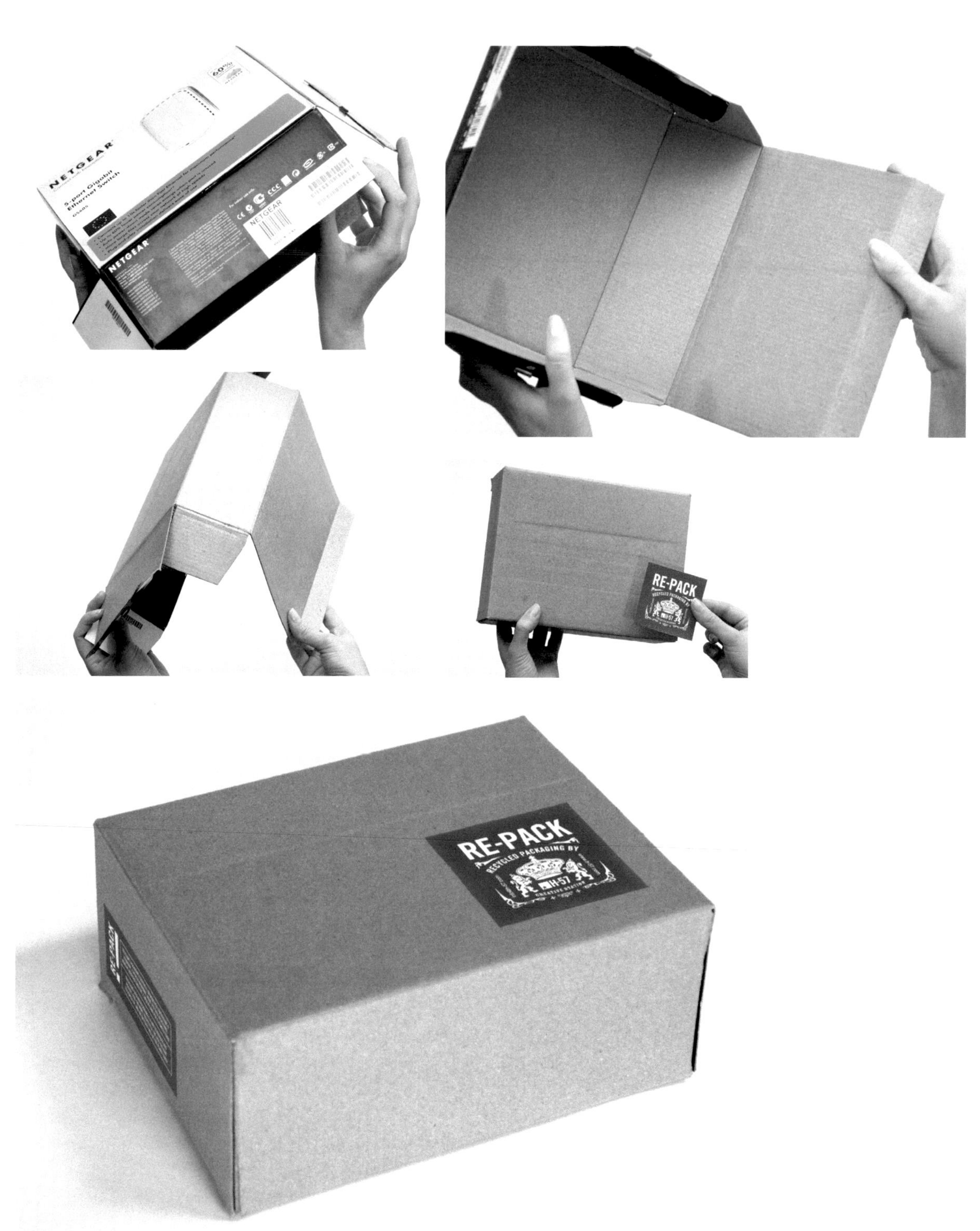

2

3

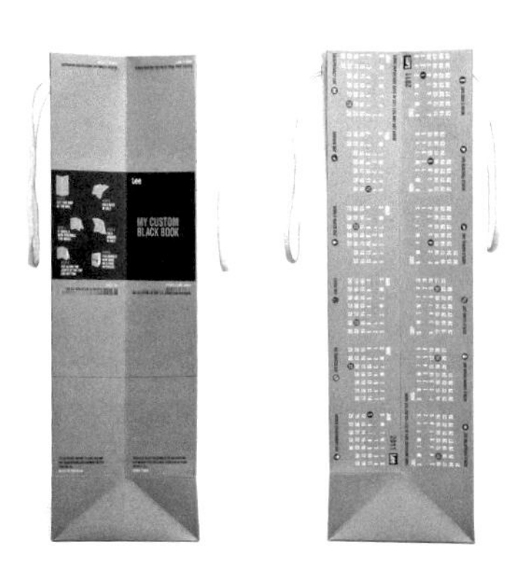

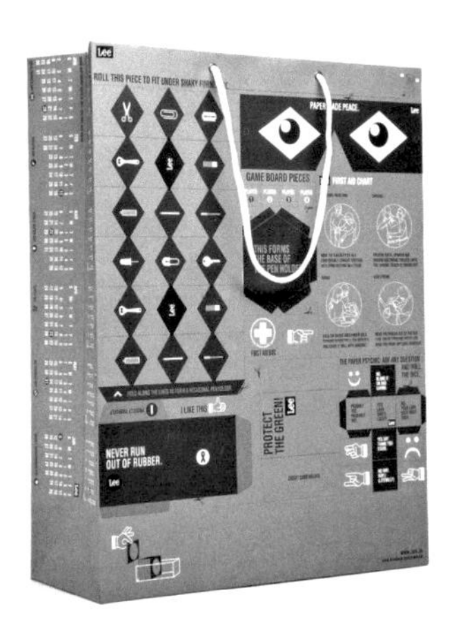
NEVER RUN
OUT OF RUBBER.

Lee
NEVER WASTED
MAKE YOUR OWN LUCK.
DO
NOT
DISTURB
ECO-FRIENDLY FUN. DO TRY THIS AT HOME.

PAPER MADE PEACE.
Lee

ECO-FRIENDLY FUN. DO TRY THIS AT HOME.

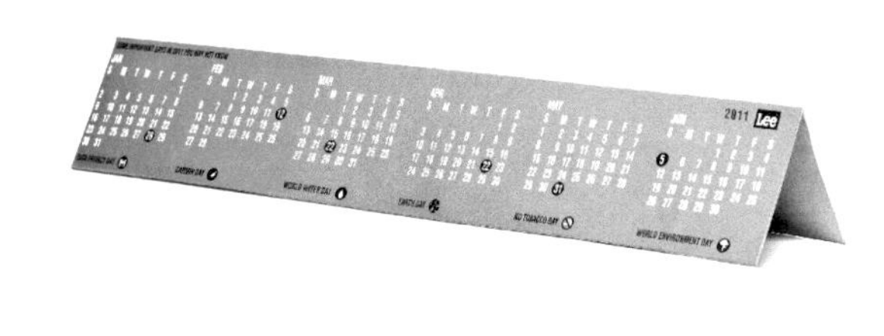
2011
Lee

4

CLEVER
LITTLE BAG

GRAPHICAL HOUSE

1 Product: Seedbom, Client: Kabloom , Distribution: UK, 2010, Material: package contains a minimum of 95% pre- and post-consumer recycled waste, produced from chlorine free (EFC) pulp, printed using vegetable based inks

NADINE BRUNET

2 Product: Pangée, student-project, 2010

Pangea is a packaging concept that combines several functions. The set of four different sized bowls are packaged in a cardboard cylinder containing edible rice that also serves to protect the bowls during transport. The bowl sizes highlight the inequity of global food distribution. While the largest bowl bears an image of North America, Africa is printed on the smallest one.

1

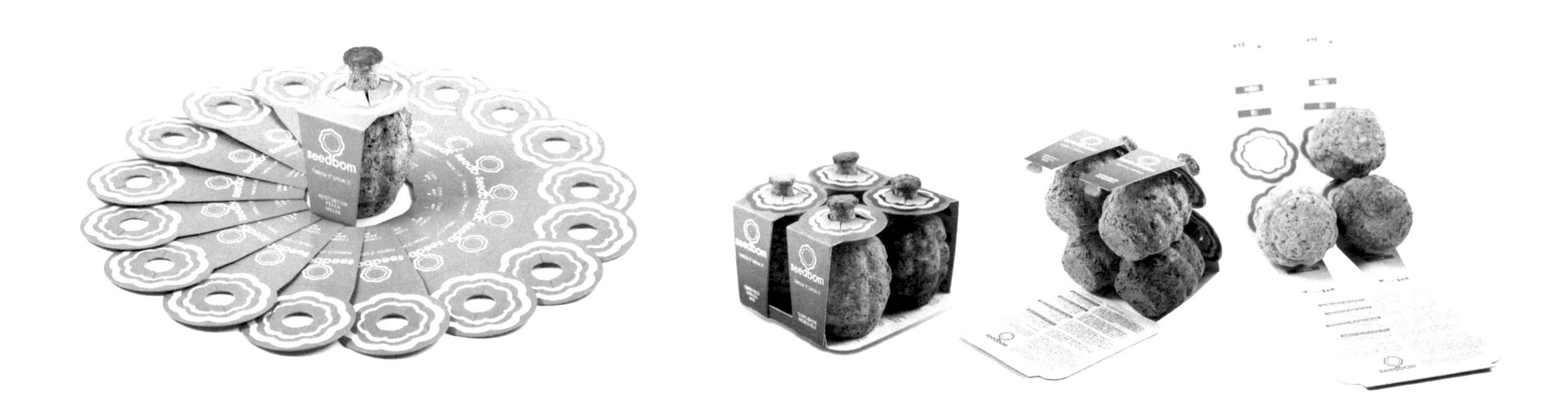

2

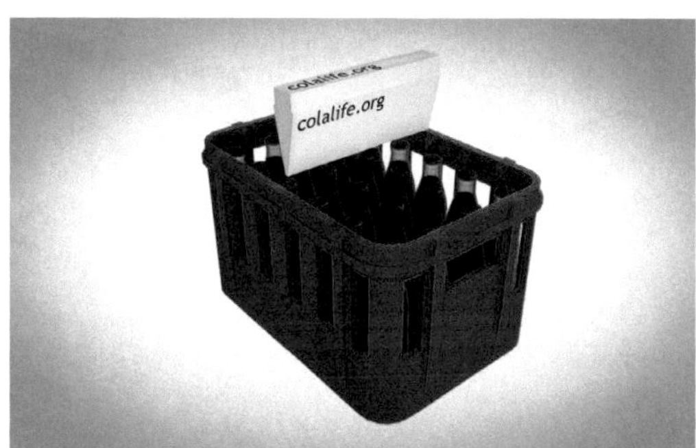

SIMON BERRY

3 Product: AidPod, Client: ColaLife, Distribution: Zambia (initially), 2008–2011

The independent non-profit organization Colalife works in developing countries to bring Coca-Cola, its bottlers, and others together to open up Coca-Cola's distribution channels to carry social products such as oral rehydration salts and zinc supplements to save children's lives. An integral part of these efforts is the development of the AidPod. The wedge-shaped pod containing the social products fits snugly into the unused space between the necks of the bottles in a Coca-Cola crate. The two-piece pod is designed with shoulders that both stabilize the packaging, and keep the pod from slipping out of the crate, while the depth of its lid is the same as its collar to resist lengthwise crushing.

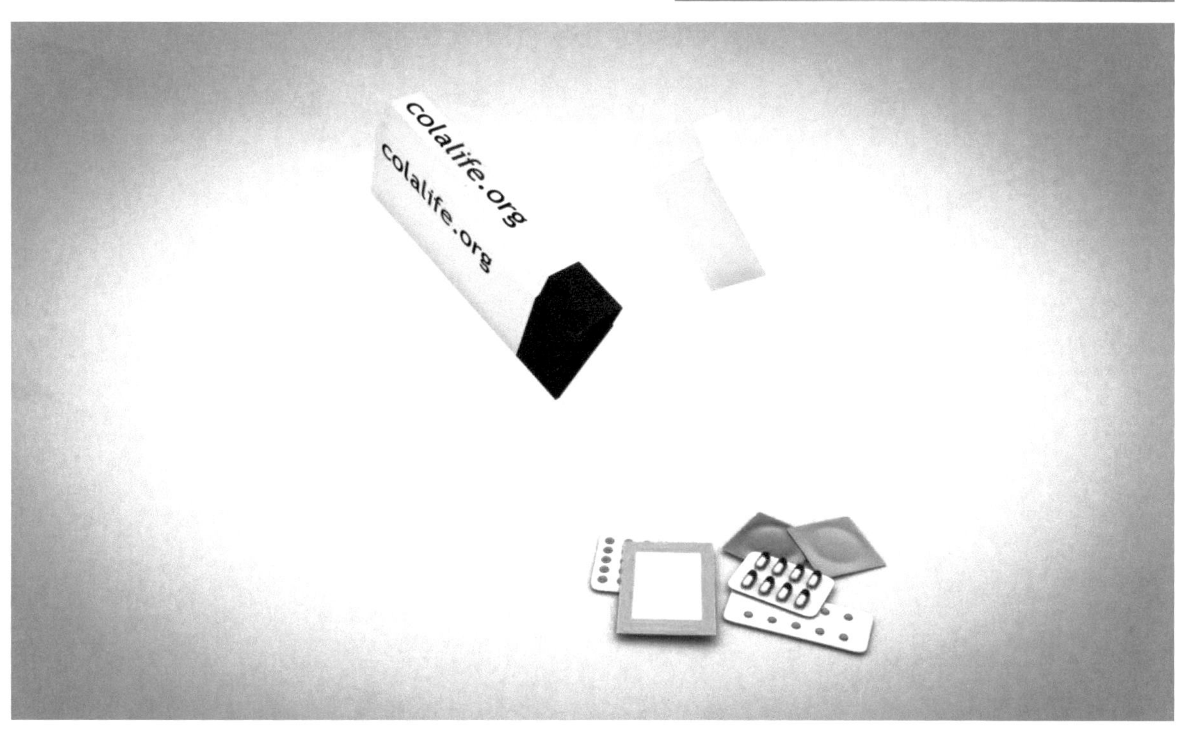

CYBERPAC

1 Product: Harmless Dissolve Magazine Mailer, Client: Creative Review, Distribution: worldwide, 2010, Material: Harmless Dissolve

The magazine packaging states explicitly what it was designed to do: "This bag dissolves in water." The Harmless Dissolve Magazine Mailer is made from a transparent, water-soluble polymer that completely biodegrades in a composting environment, in a dishwasher, or in a washing machine.

ELEVEN1.COM

2 Client: Replenish, Distribution: USA, 2010

Why pay for water and plastic? In a normal bottle of household cleaner, only 5% is cleaner, while the rest is plastic and water; in contrast, Replenish is designed to save the consumer money while using 90% less plastic, oil and CO_2 emissions than its conventional competitors. The eco-friendly, multi-surface household cleaner consists of a reusable, recyclable PET spray bottle with built-in measuring cup and a replaceable pod filled with concentrated cleaner. Flip the bottle upside-down and squeeze the measured amount of concentrate from the pod, mix in your own water from home and clean.

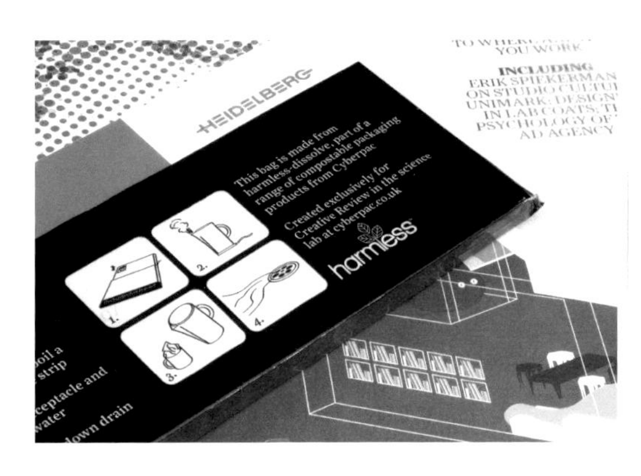

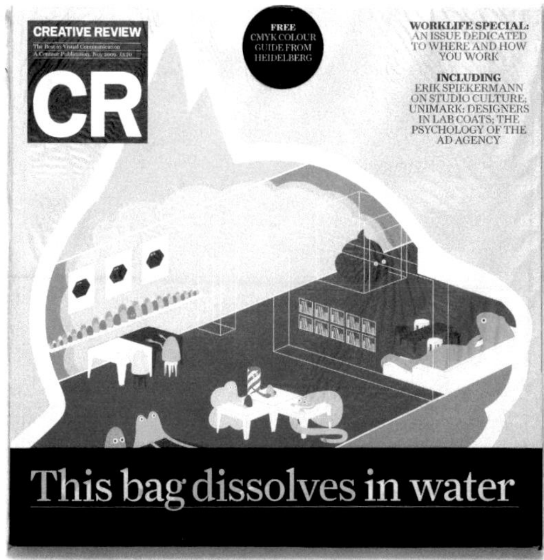

1

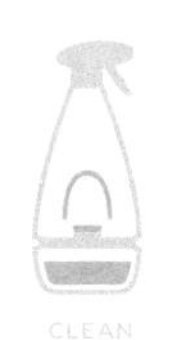

2

3

KOREFE

3 Product: Stop the Water While Using Me!, Client: T.D.G. Vertriebs GmbH & Co. KG, Distribution: Germany, Austria, 2010, Material: recyclable plastic

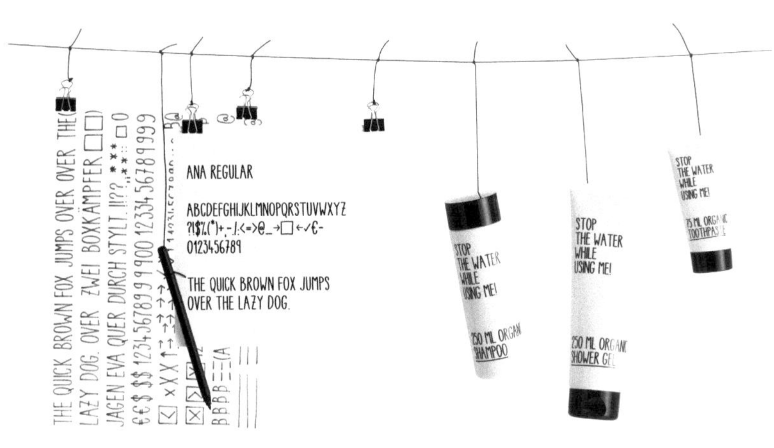

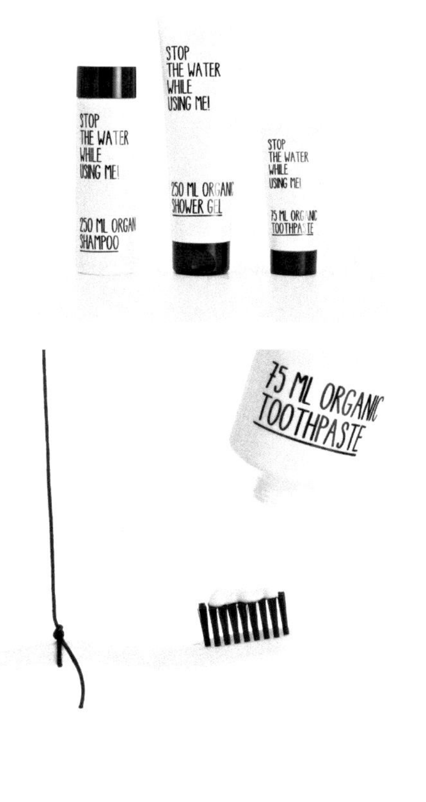

SENTRAL DESIGN STUDIO, & ÅSA JUNGNELIUS KOSTA BODA
Client: Agonist Parfums, Distribution: worldwide, 2008–2010

AGONIST

NEAL FLETCHER

1 Product: Six Servings of Italian Spaghetti, self-initiated, Distribution: UK, 2010, Material: cardboard, glue
Assuming that a standard 500 g pack of spaghetti yields approximately six servings, this refillable spaghetti box is designed to hold six servings of spaghetti in separate compartments.

STÉPHANIE SANSREGRET

2 Product: L'Kit de Survie, student-project, 2011

SABRINA LÉVESQUE

3 Product: Bulbo, student-project, 2010, Material: thermoplastic and paper
A variety of pesto flavors is packaged in individual portions together in a bouquet evoking the shape of a garlic bulb.

1

2

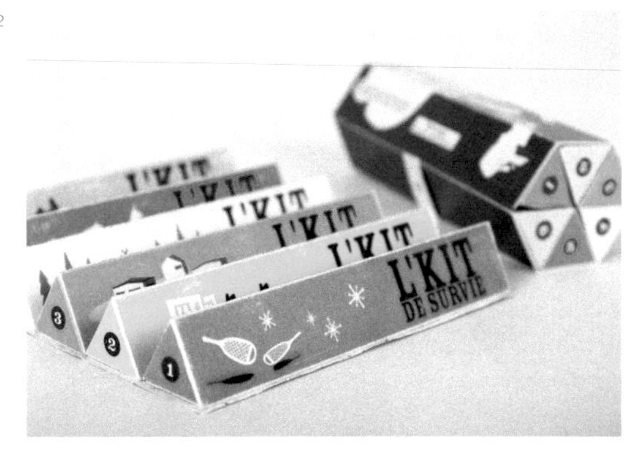

3

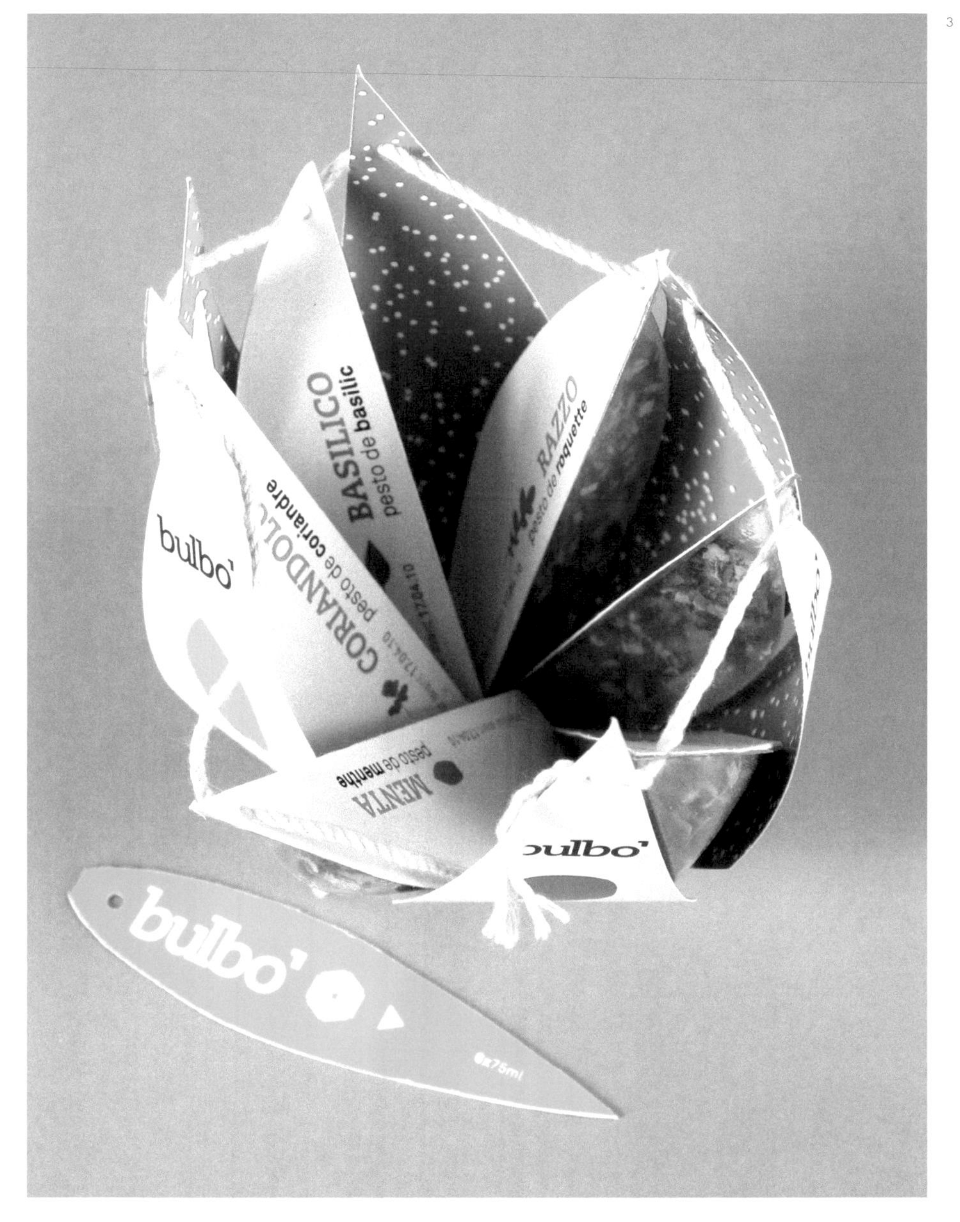

4

5

6

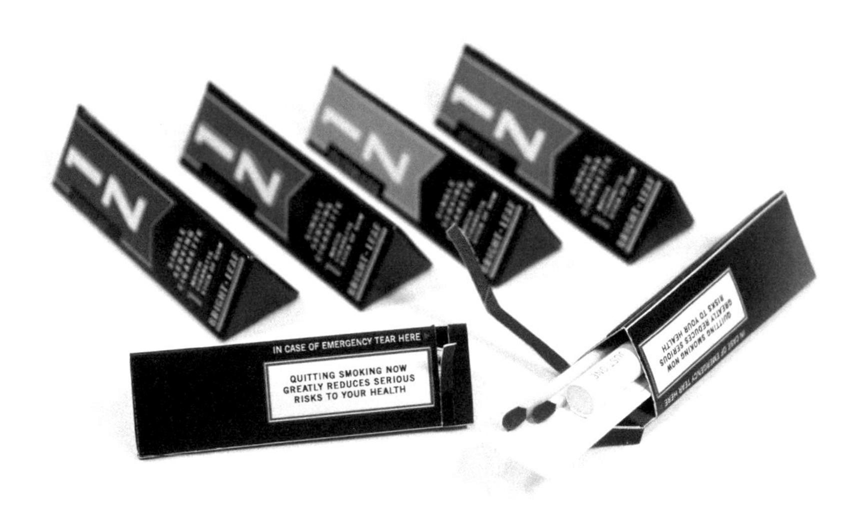

REYNOLDS AND REYNER

4 Product: Antismoke pack, 2010

UGO VARIN

5 Product: Cancer & Deuil, student-project, 2011

CONTINUUM

6 Product: 1Z Single-serve Cigarette, self-initiated, 2010, Material: laser-printed cardstock, 2 matches, 1 cigarette, striking surface

A range of single-serve cigarettes offers an emergency fallback in various tobacco flavors and strengths for the smoker trying to quit, eliminating the need to purchase an entire pack.

JOY LIN

1 Product: Crème Brûlée Set (whisk, butane hand torch, 2 oz. ramekins), Client: Envie, 2010, Material: red oak and PETG
The gourmet gift set for the preparation of crème brûlée presents its product as an experience to be envied. Cooking utensils are contained in a red oak trapezoid whose windows offer an enticing glimpse of the contents through the package's layers. The exquisitely designed hinged box pivots outwards, creating a unique opening ceremony.

PETAR PAVLOV

2 Product: Doritos, self-initiated, 2009
A stay-fresh packaging concept for Doritos chips inspired by the chips' own shape. The packaging's triangular structure enables it to be reclosed after opening.

BRYANT YEE

3 Inside and Outside the Box: Redesigning LED Packaging, Senior Thesis Project, 2011
Thinking outside as well as inside the box was the aim of this thesis project on socially responsible LED packaging. The stackable boxes make clever use of post-consumer recycled paper, lucid graphics and succinct copy in eye-catching as well as informative design. The packaging can be reused as a mailer in order to recycle the product.

2

TOASTED CORN
Doritos
TOASTED CORN
Doritos
TOASTED CORN
TOASTED CORN
Doritos
TOASTED CORN
Doritos

NACHO CHEESE
Doritos
NACHO CHEESE
Doritos
NACHO CHEESE
Doritos

3

PHILIPSLED
WHITE
Brightness
450
LUMENS
PHILIPSLED
COOL WHITE
Brightness
450
LUMENS
Life
25,000
HOURS
PHILIPSLED
WARM WHITE
Brightness
450
LUMENS
PHILIPSLED
WHITE
Brightness
800
LUMENS
PHILIPSLED
COOL WHITE
Brightness
800
LUMENS
PHILIPSLED
WARM WHITE
Brightness
800
LUMENS

POSTAGE
WILL BE
PAID BY
ADDRESSEE
Recycle Ann Arbor
2420 South Industrial Highway
Ann Arbor, MI 48104

SERGE RHÉAUME

1 Product: Vetro, student-project, 2011, Material: multiply board, and light adhesive stripe of paper

A design project demonstrating that chic must not be a matter of excess: four cocktail glasses are held together by a simple band of paper, and a separator that can be disassembled and used as coasters.

DAVID THÉROUX

2 Product: Collet, student-project, 2011, Material: kraftboard

A single piece of recycled, and recyclable cardboard requiring neither ink nor gluing serves as the sole material of this environmentally conscious packaging design for a beer six-pack. Product information is printed directly onto the bottles, and the base of the cardboard packaging can be broken up into square coasters.

1

2

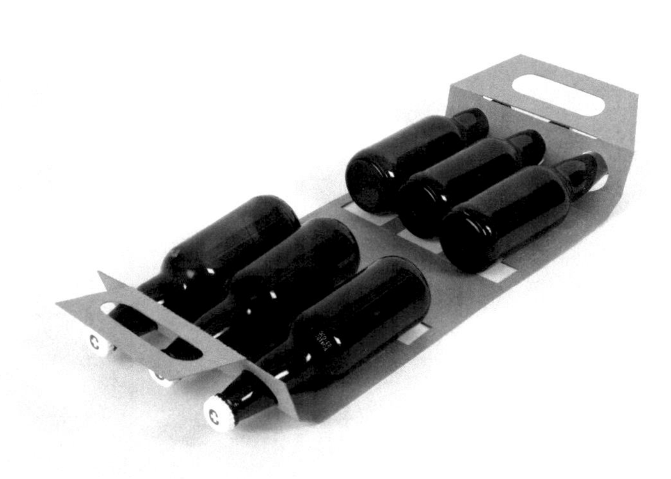

3

NICOLAS MÉNARD

3 Product: Antidote, student-project, 2011, Material: kraftboard

In times when the computer savvy increasingly obtain their software via digital download, a market still seems to remain for those who appreciate a physical presentation of their product. This sustainably-minded concept positions itself in this niche by doing away with extraneous content for effective communication, minimizing waste, and efficient transport. The front label identifies the product, and seals the slim and simple cardboard envelope, which opens up to reveal its components. A reusable USB key holds the installation software and users' manual. The serial number, and other information are presented on a business-size card that can be easily stored in standard business card holders.

IVAN MAXIMOV

4 Product: MUG Beer, Client: MUG Pub, student-project at British Higher School of Art and Design (Moscow), Russia, 2011, Material: paper, paperboard, cardboard

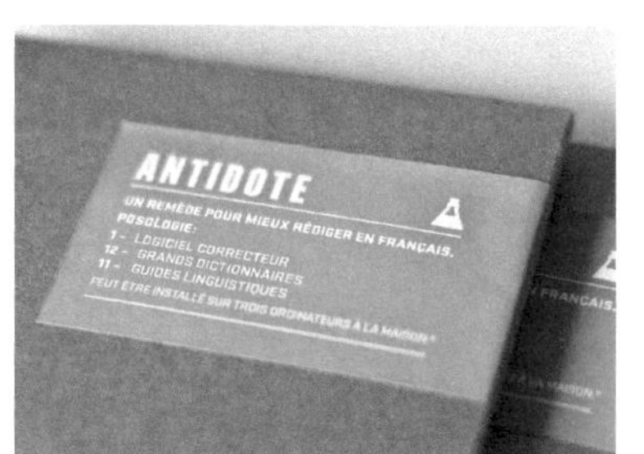

4

THE.
Mihoko Ouchi and Sherwood Forlee
1 Product: Anti-Theft Lunch Bags, worldwide, 2009, Material: food-safe reusable and recyclable LDPE

SCOTT AMRON
2 Product: Heatswell Coffee Cup
Prototype for a stackable, disposable coffee cup with a heat activated insulation sleeve.

TO-GENKYO
Naoki Hirota,Yuki Ijiri,Koji Takahashi
3 Product: Fresh Label, concept, 2008
False labelling on food is a worldwide concern. The proposed food label changes color in reaction to ammonia emitted by food as it becomes spoiled. When it is no longer edible, the label darkens to reveal a pattern that renders the barcode unscannable. Designed in the shape of an hourglass, the label instantly conveys its purpose to consumers. This active visualization of the product's shelf life creates a new relationship between consumers and comestibles.

SCHOLZ & FRIENDS BERLIN
4 Product: Festina Profundo, Client: Festina, Distribution: PET/PE

1

2

3

牛ローススライス
原材料名/牛（国産）
399
200
798

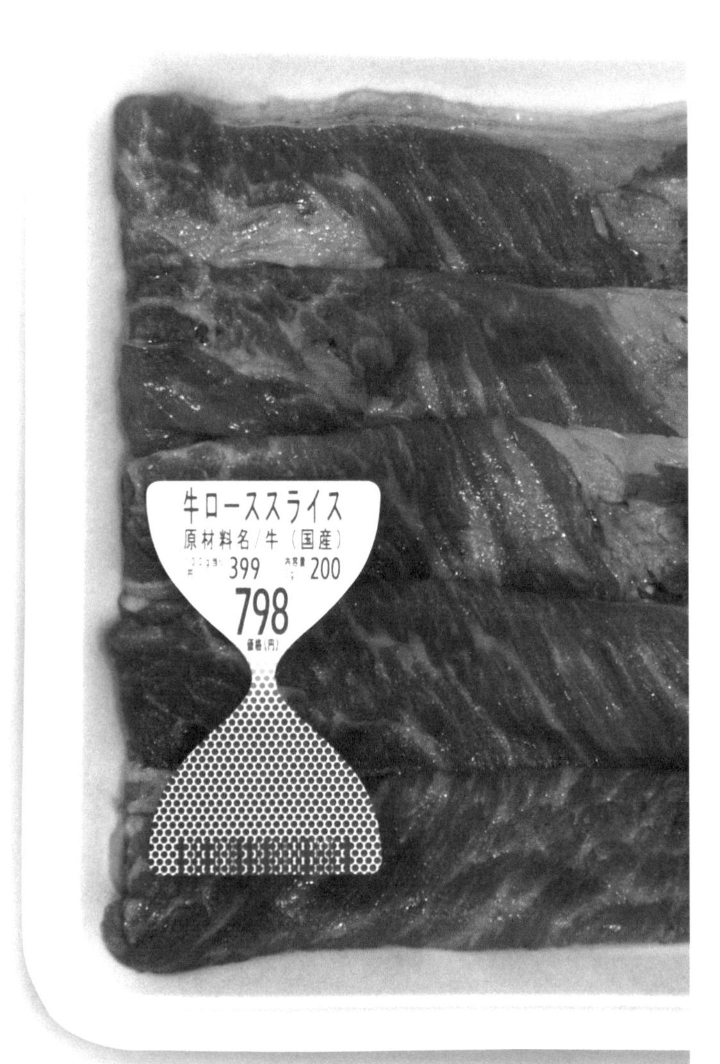

牛ローススライス
原材料名/牛（国産）
399
200
798

4

FESTINA
FESTINA
engineered for water
FESTINA
engineered for water

STEPHANIE KUGA

1 Product: The Gift of Life, self-initiated, Distribution: Concept, 2009, Material: e-flute cardboard

Potential organ donors are encouraged to give the gift of life with this range of promotional gift boxes that open to reveal a sewn cloth liver, lungs, heart, kidney, etc., and information on the incentives for organ donation.

1

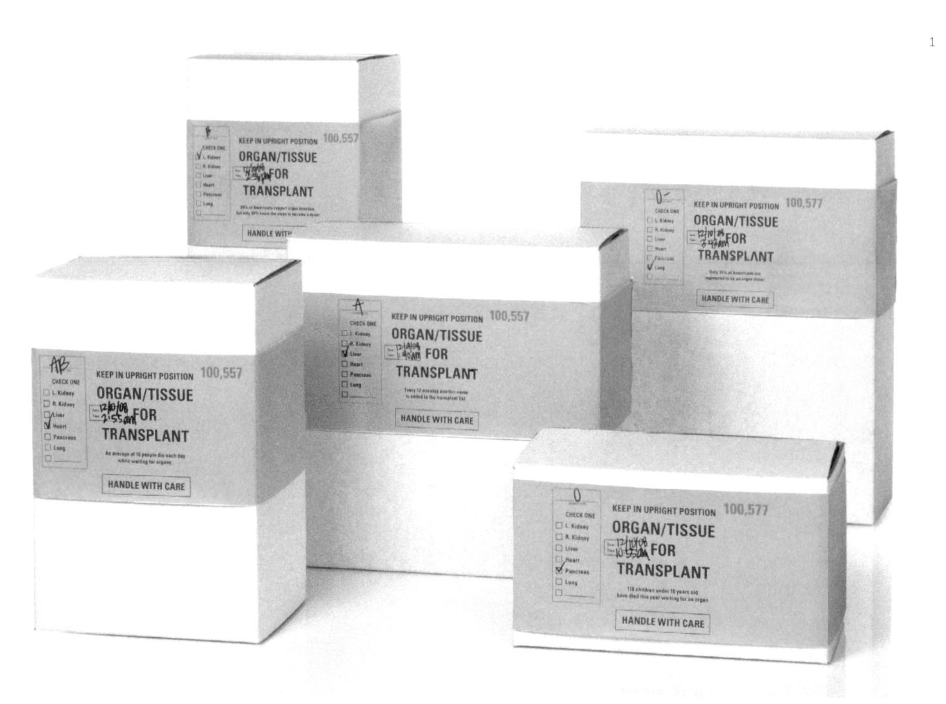

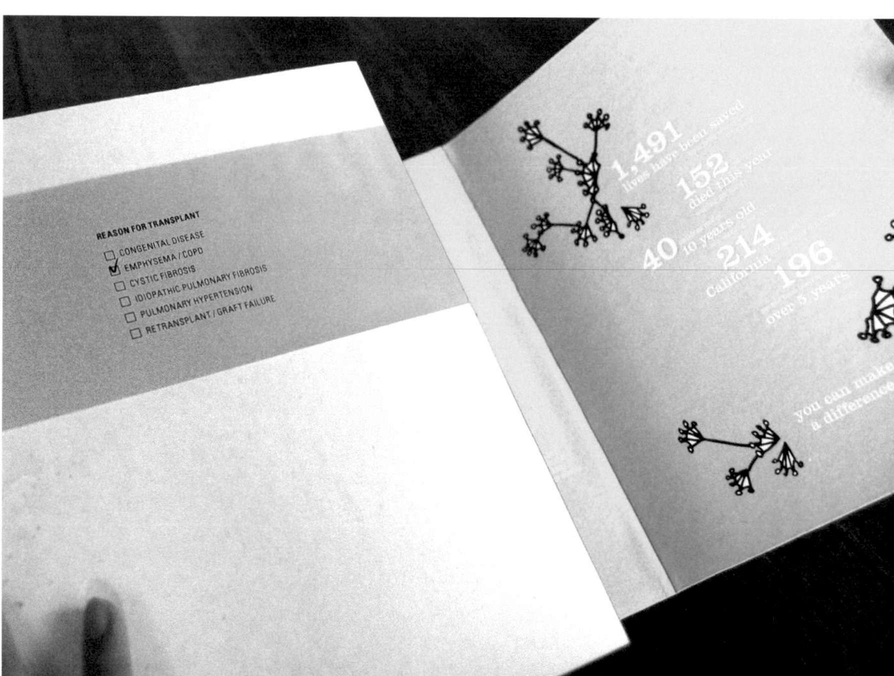

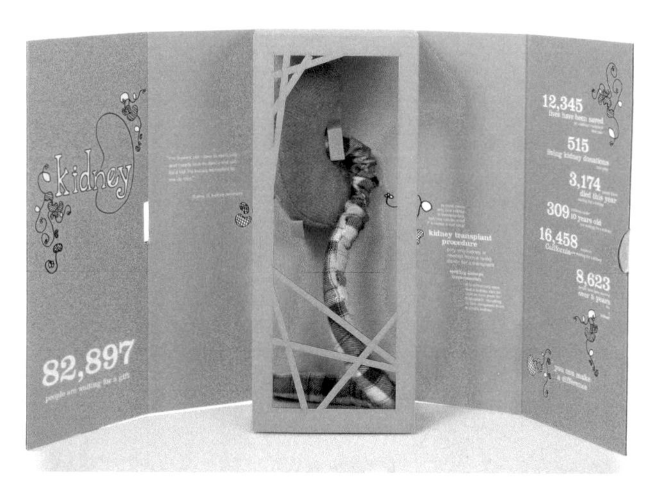

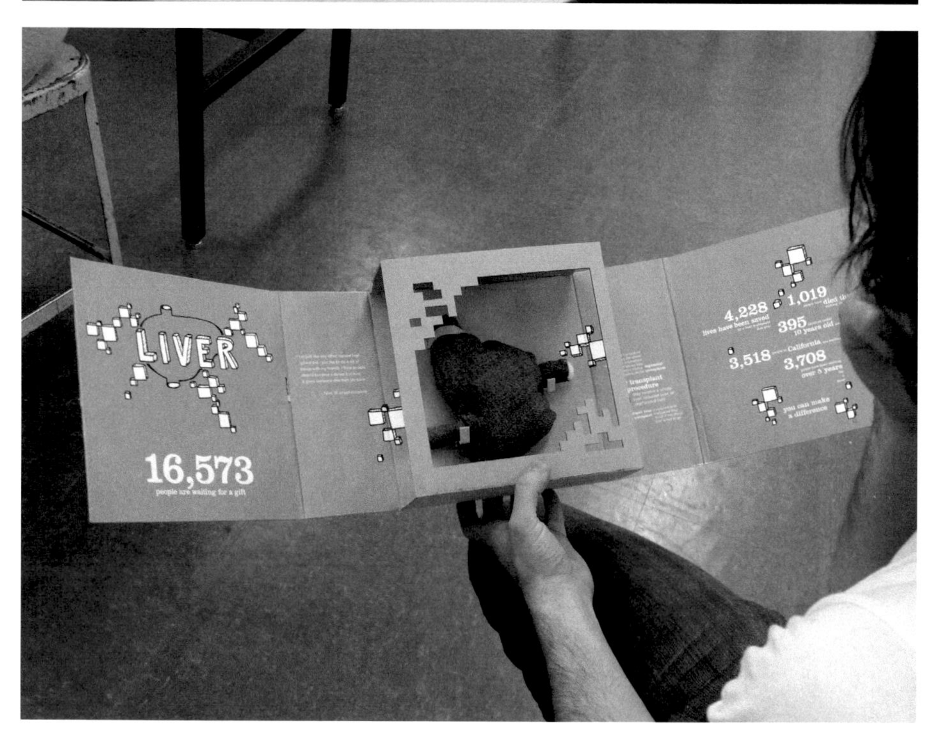

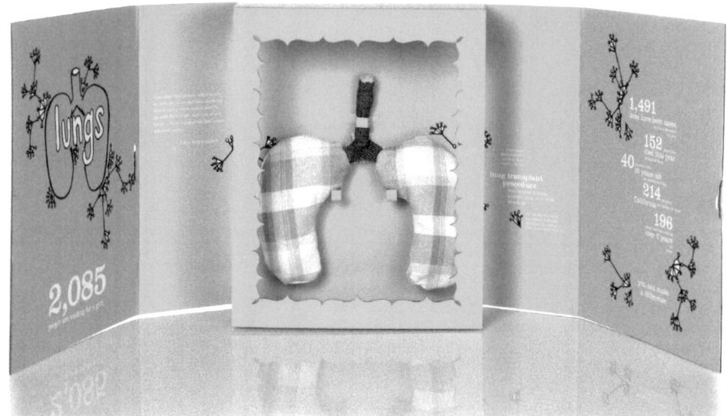

BUREAU BRUNEAU

[2] Product: The Federal Emergency Management Agency, self-initiated, 2010, Material: vintage naturally sunburst papers found in grandmothers basement

For the designer's final graduation exam, a comprehensive project was built around the premise of a post-apocalyptic, and dystopic world. Here, the packaging for survival rations produced by a futuristic manifestation of the U.S. Federal Emergency Management Agency (FEMA) emphasizes informative and clear communication with the aim of fostering the reconstruction of a new global society.

SIMON L'ARCHEVÊQUE

1 Product: Mon lait, student-project, 2011
Proposing a new approach to packaging milk, this student project establishes a link between the heritage of Quebec's milk and the development of traditional Canadian housing—from the First Nations to modernity.

PACKLAB PARTNERS

2 Product: Al Rawabi's Family Milk Packaging Range, Client: Al Rawabi Dairy Company L.L.C, Distribution: UAE, 2010
The milk bottle designed for this local dairy company gives a unique identity to the product not only through its iconic form, but also purports to improve usability in comparison to its competitors.

AUDREE LAPIERRE

3 Product: Nutritional Facts, self-initiated, 2010
In this self-initiated project, graphic visualisations of milk's nutritional data are used to brand the milk cartons—not to mention providing useful information about the product, such as caloric ratio, nutrient balance, and amounts of carbohydrates, fat, protein and sodium per serving.

VISUALDEVICE

4 Product: Milk Packaging
An experiment in cardboard packaging as well as in communication, this project represents the content of the product in the simplest way, both through word and image. The packaging bears the same dimensions as a two-liter carton of milk.

1

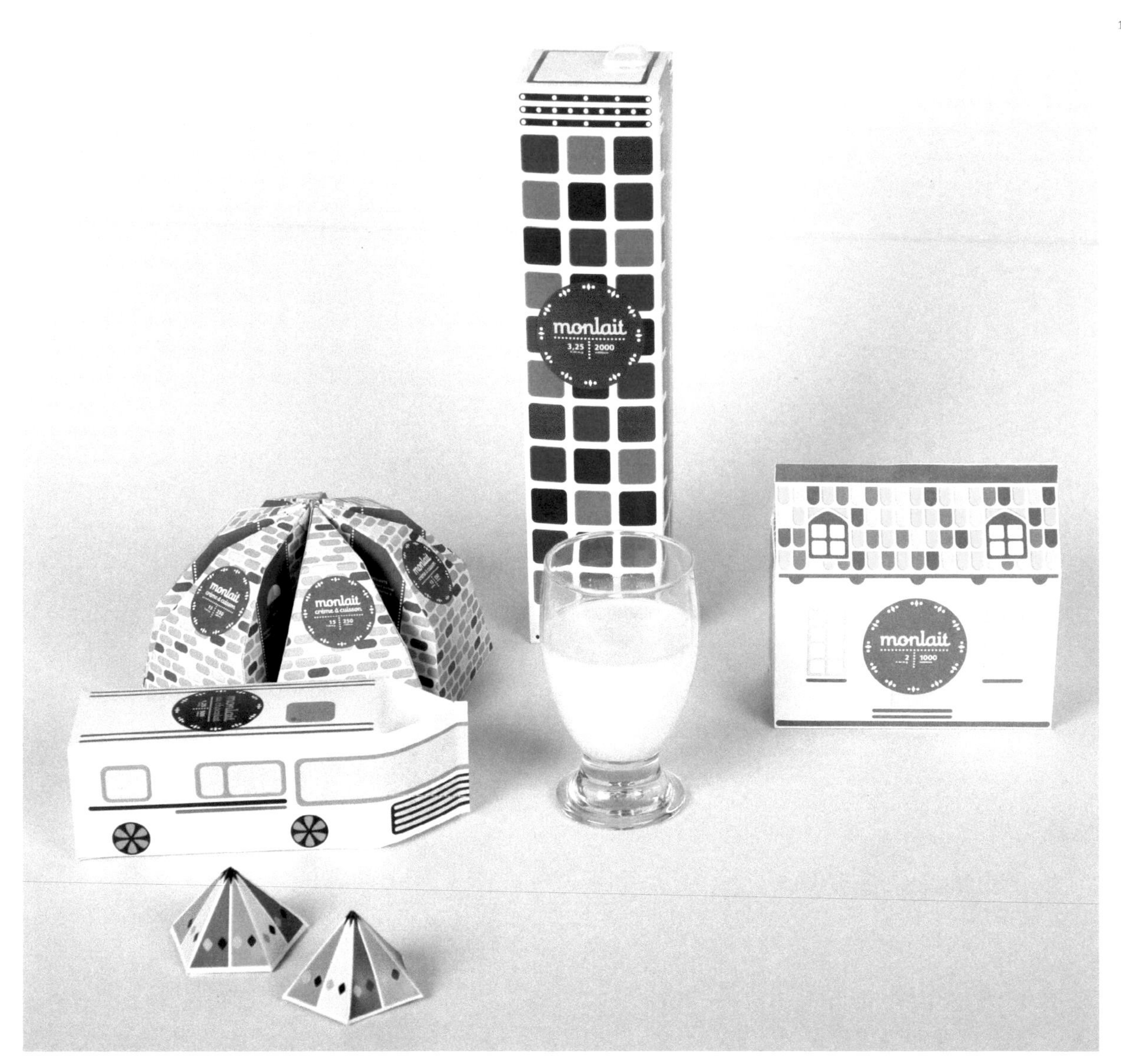

2

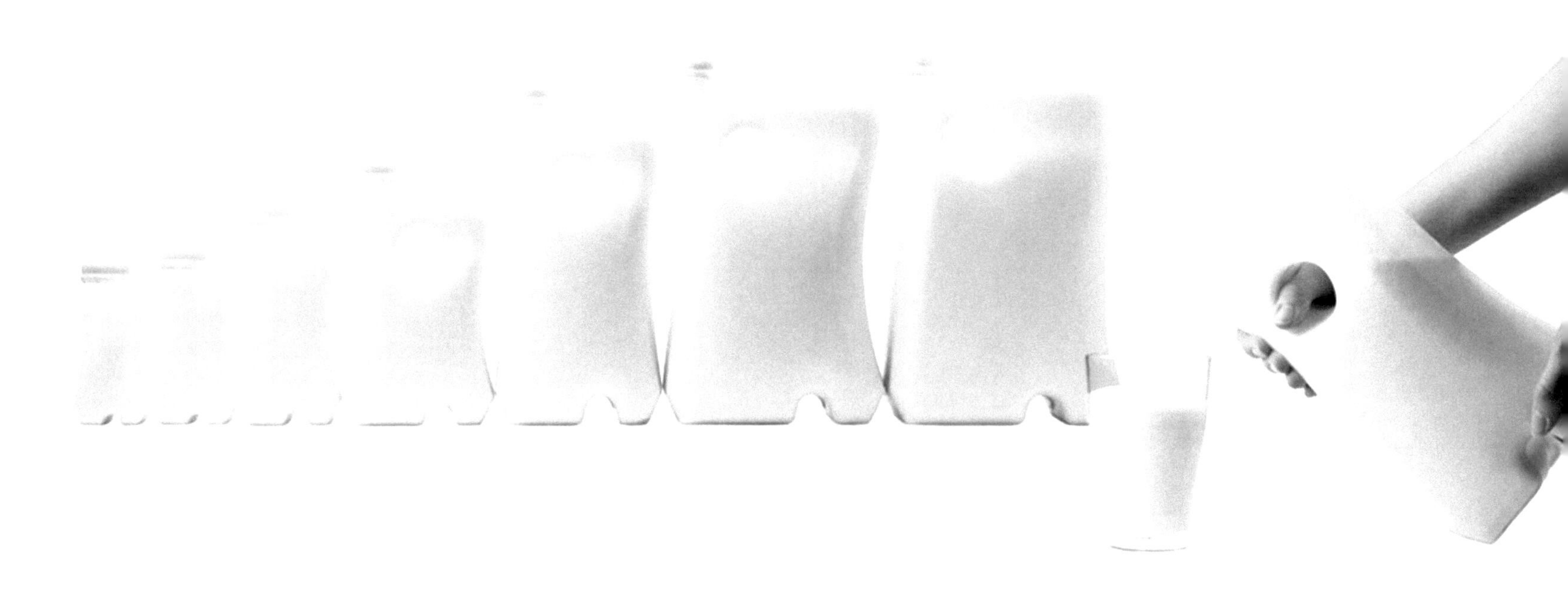

3

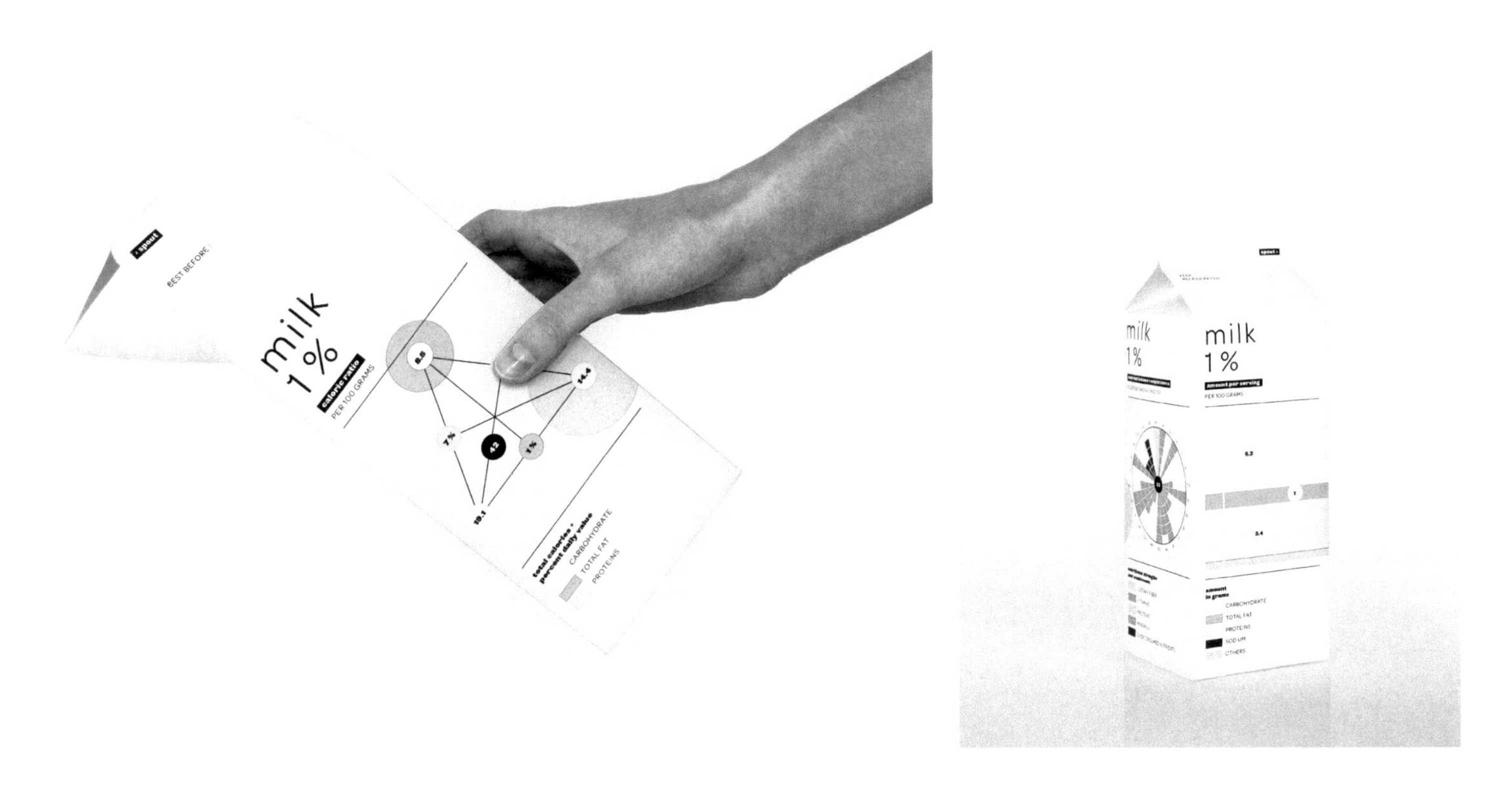

4

1-9

B

C

ÉVA VALICSEK
Hungary
evivalicsek.tumblr.com
p. 148

EVERYTHING
New Zealand
www.everythingdesign.co.nz
p. 22

EYAL BAUMERT
Israel
www.eyalbaumert.com
p. 125

FABIO MILITO DESIGN
Fabio Milito, Francesca Guidotti
Italy
www.fabiomilito.com
pp. 108–109

FRAN ALDEA
Chile
franaldea.blogspot.com
p. 24

FRANK ALOI
Australia
www.frankaloi.com.au
p. 136

FREDDY TAYLOR
United Kingdom
www.freddytaylor.co.uk
p. 24

GLASFURD & WALKER DESIGN
Canada
www.glasfurdandwalker.com
p. 129

GOODMORNING TECHNOLOGY
Denmark
www.gmtn.dk
p. 124

GRAFIKART
Edward Pearson
Chile
www.grafikart.cl
p. 143
Naming: Sofia Gallinal, Photography: Luis Piano

GRAPHICAL HOUSE
United Kingdom
www.kabloom.co.uk
p. 200
Photography: Susan Castillo

H-57 CREATIVE STATION
Italy
www.h-57.com
pp. 184, 198

HAPPY CREATIVE SERVICES (INDIA) PVT LTD
India
www.thinkhappy.biz
p. 199
Art Direction: Viduthalai Raj, Creative Direction: Praveen Das, Kartik Iyer, Copywriting: Athul CT, Client Relations: Neelima Kariappa

HASAN & PARTNERS
Finland
hasanpartners.net
p. 19

HATCH DESIGN
USA
www.hatchsf.com
p. 157
Art Direction: Katie Jain & Joel Templin, Design: Shadi Kashefi
p. 182
Art Direction: Joel Templin & Katie Jain

HELEN MARIA BÄCKSTRÖM
Sweden
helenmariabackstrom.se
p. 17

HELVETICA INC.
Japan
www.helvetica.co.jp
p. 159

HERE DESIGN
United Kingdom
www.heredesign.co.uk
p. 141

HINOMOTO DESIGN
Shigeki Kunimatsu
Japan
www.hinomoto-d.com
p. 129

IDENTITYWORKS
Henrik Hallberg
Sweden
www.identityworks.se
p. 17
Account Direction: Anette Brolin Berg, Creative Direction: Nils Jensen

IGOR MANASTERIOTTI & MIA MARIC
Croatia
www.manasteriotti.com
p. 11

ILOVEDUST
United Kingdom
www.ilovedust.com
pp. 36–37, 65

IRVING & CO
Julian Roberts
United Kingdom
www.irvingandco.com
p. 127
p. 162
Illustration: Valérie Roy
p. 163
Illustration: Lucia Gaggioti

IVAN MAXIMOV
Russia
cargocollective.com/maximov
p. 211

JDA, INC. RETAIL READY DESIGN
USA
www.jdainc.com
p. 10
Creative Direction: David Jensen, Art Direction: Jerome Calleja, Design: Dean Kojima, Stephanie Han

JESSIE WHIPPLE VICKERY
USA
www.kunstkraft.com
p. 169
Photography courtesy of Stumptown

JONATHAN ADLER
USA
www.jonathanadler.com
p. 160

JONES KNOWLES RITCHIE
United Kingdom
www.jkr.co.uk
p. 60
p. 160
Photography: Andy Grimshaw

JOY LIN
USA
www.thejoylin.com
p. 208
Guidance: Dan Hoy

JULIAN BAKER
USA
theywillseeme.com
p. 133

JUST BE NICE STUDIO
Russia
www.justbenicestudio.com
p. 50

KIMBERLY-CLARK DESIGN
USA
www.kimberly-clark.com
pp. 48–49
Creative Direction: Christine Mau Design, Management: Jody Douglas, Art Direction: Jennifer Brock / Katie Hinrichs, Illustration: Hiroko Sanders

KOREFE
Germany
www.korefe.de/
pp. 80–83
Creative Direction: Katrin Oeding
Art Direction: Reginald Wagner, Jan Simmerl, Jan Hartwig, Text: Till Felber, Fabian Bill, Gereon Klug, Thomas Völker, Katharina Trumbach, Madelen Gwosdz, Photography: Ulrike Kirmse
p. 203
Creative Direction: Katrin Oeding
Art Direction: Christian Doering, Ana Magalhães, Christine Knies, Text: Till Felber, Thomas Völker, Gereon Klug, Photography: Robin Kranz

KYLE TEZAK
USA
www.kyletezak.com
p. 178
Illustration: Gil Elvgren

LAINEY LEE
USA
www.laineylee.com
p. 96
Art Direction: Kristin Sommese

LAURA BERGLUND
USA
lauraberglund.com, design-ranch.com
p. 28
Professor: Michael Kidwell, School: Kansas City Art Institute

LINDSEY FAYE SHERMAN
USA
www.lindseyfaye.com
p. 50

LINNA XU
Canada
linnaxu.com
pp. 190–191

LITTLE FURY
USA
www.littlefury.com
p. 107

LOUISE FILI
USA
www.louisefili.com
pp. 158–159

LOWE BRINDFORS
Sweden
www.lebureau.se
p. 19
Art Direction: Lina Elfstrand, Noel Pretorius & Pia-Maria Falk, Illustration: Sac Magique, Final Art: Mimmi Wägerth

LUIS TRÉPANIER
Canada
www.packaginguqam.blogspot.com
p. 65
Designed at Packaging UQAM / Sylvain Allard (Professor)

MADS JAKOB POULSEN
USA
www.madsjakobpoulsen.dk
p. 36

MANUAL
USA
www.manualcreative.com
p. 92
Creative Direction: Tom Crabtree, Design: Tom Crabtree, Joshua Swanbeck, Eileen Lee, Production: Uneka Concepts

MARC L. COOPER
USA
www.coopsmicrocreamery.com
p. 166

S

U–Z

BOXED & LABELLED

* NEW APPROACHES TO PACKAGING DESIGN *

TWO!

Edited by Robert Klanten, Sven Ehmann
Text and preface by Sylvain Allard

Cover by Matthias Hübner for Gestalten
Cover photography by Hatch Design
Layout by Matthias Hübner for Gestalten

Project management by Vanessa Diehl for Gestalten
Production management by Martin Bretschneider for Gestalten
Copyediting by Alisa Lieu Kotmair
Proofreading by Leina Gonzalez
Printed by Offsetdruckerei Grammlich, Pliezhausen
Made in Germany

Published by Gestalten, Berlin 2011
ISBN 978-3-89955-378-9

For more information, please visit www.gestalten.com.

Bibliographic information published by the Deutsche Nationalbibliothek.

The Deutsche Nationalbibliothek lists this publication in the Deutsche National-bibliografie; detailed bibliographic data are available online at http://dnb.d-nb.de.

None of the content in this book was published in exchange for payment by commercial parties or designers; Gestalten selected all included work based solely on its artistic merit.

This book was printed according to the internationally accepted ISO 14001 standards for environmental protection, which specify requirements for an environmental management system.

This book was printed on paper certified by the FSC®.

Gestalten is a climate-neutral company. We collaborate with the non-profit carbon offset provider myclimate (www.myclimate.org) to neutralize the company's carbon footprint produced through our worldwide business activities by investing in projects that reduce CO_2 emissions (www.gestalten.com/myclimate).